The Lead Dog Has the Best View

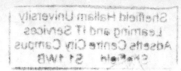

Other Titles of Interest

Civil Engineering Practice in the Twenty-First Century:
Knowledge and Skills for Design and Management.
2001. ISBN 0-7844-0526-3.

Engineering Your Future, Second Edition: The Non-Technical Side
of Professional Practice in Engineering and Other Technical Fields.
2000. ISBN 0-7844-0489-5.

Excellent Communication Skills Required for Engineering Managers.
1994. ISBN 0-7844-0047-4.

How to Produce Effective Operations and Maintenance Manuals.
2000. ISBN 0-7844-0011-3.

How to Work Effectively with Consulting Engineers:
Getting the Best Project at the Right Price
(ASCE Manuals and Reports on Engineering Practice No. 45, revised edition).
2003. ISBN 0-7844-0637-5.

Managing and Leading: 52 Lessons Learned for Engineers.
2004. ISBN 0-7844-0675-8.

Personal Success Strategies: Developing Your Potential.
1999. ISBN 0-7844-0446-1.

The Lead Dog
Has the Best View

Leading Your Project Team to Success

Gordon Culp, P.E.

Anne Smith, P.E.

Library of Congress Cataloging-in-Publication Data

Culp, Gordon L.
   The lead dog has the best view : leading your project team to success. / Gordon Culp,
      Anne Smith.
      p.   cm.
   Includes bibliographical references and index.
   ISBN 0-7844-0757-6 (alk. paper)
   1. Project management. 2. Leadership. I. Smith, R. Anne. II. Title.
   HD69.P75C84   2004
   658.4'092—dc22

                         2004026341

Published by American Society of Civil Engineers
1801 Alexander Bell Drive
Reston, Virginia 20191
www.asce pubs.asce.org

Contents

Figures

Preface

The Lead Dog

While on a winter trip to Alaska, we happened upon a local park where a group of dogsled racers was gathered for some weekend practice. Having never seen dogsleds in action, we were fascinated to watch the teams preparing. As the individual dogs were being placed in the team harness, the drivers tied the dogsleds to poles in the ground to keep the dog teams from racing off. One of the teams accidentally came untied, and the team went racing down the trail toward the finish line with the driver in hot pursuit on foot. It later occurred to us that we had witnessed a demonstration of the energy that is released when teams are led and not merely managed.

A controlling project manager's style often constrains the team, like tying the team to a pole. Certainly the management functions of seeing that the team had the right equipment and provisions for the long race were critical, but it was when the dog team broke loose from the pole and its leader—the lead dog—took over that the energy of the team was released. The team raced ahead, following their very enthusiastic leader, who had a clear vision of the objective and the path to that objective. The entire team was caught up in the leader's enthusiasm and excitement.

Wouldn't it be wonderful if the typical project team had this kind of enthusiasm and immediately raced at full speed in a direct line toward the project objectives? When project teams are led rather than merely managed, we see similar energy and enthusiasm. Not only is the project team more excited when led, but by advancing from being a manager to being a leader, the project leader has a more fulfilling and rewarding experience—after all, the lead dog definitely has the best view.

Purpose of this Book

We have written this book to describe what it takes to make the transition from project manager to project leader. Too many project managers rely on their technical skills and computer-generated information about project status in their attempts to achieve project success. They confuse management with leadership. It is rare that a project fails because of the lack of technical skills or the inability to draw a critical path schedule. It is more common for projects to fail because of relationship and communication difficulties related to poor project leadership. The tremendous natural energy of team members is unleashed when they are led rather than merely managed.

When led, teams have more energy, are more enthusiastic, are more effective, and succeed more often. Just like the dog team we saw, the teams are energized and moving at top speed in the right direction toward a common vision that constitutes success. Leaders provide motivating environments that allow the tremendous synergy of the team's individual talents and energy to occur.

It is our goal for this book to provide a person-centered, results-driven approach to project leadership that is not found elsewhere in the literature. We have included some management tools such as checklists to facilitate management tasks so that leaders can focus on leadership.

How to Use this Book to Improve Your Leadership Skills

This book is intended to be a frequently used tool, not read once and placed on a shelf. Each chapter includes exercises that you can use on an ongoing basis to periodically assess and improve your leadership skills. Some of the exercises can be used with your team, while others are for your individual use. The team exercises have the added benefit of creating a team-building opportunity. The book contains many checklists that you can use on an ongoing basis to quickly recall and apply the key concepts on your projects. Our hope is that the book will be a tool that makes your project work more fulfilling, both professionally and personally.

Gordon Culp, P.E., and Anne Smith, P.E.
Smith Culp Consulting
Las Vegas, Nevada
www.smithculp.com

About the Authors

Gordon Culp, P.E., and Anne Smith, P.E., have a combined 60 years of experience in leading projects with costs ranging from a few thousand to millions of dollars. Both have degrees in engineering and applied psychology.

Prior to the formation of Smith Culp Consulting in 1992, Gordon Culp led the engineering firm of Culp, Wesner, Culp until it was acquired by a large, national engineering firm. He then served as Executive Vice President of the 1000-person national engineering firm.

Anne Smith, after being a project manager for several years, served as the national director of training for a large architectural/engineering firm with 23 offices and a staff of 1400.

As the principals of Smith Culp Consulting, they provide management consulting, training, and facilitation services to a wide range of public and private clients. They have conducted leadership and project management trainings for ASCE and other professional societies. Their experiences from applying their unique combination of engineering-based insights into the practical aspects of getting projects done and psychology-based insights into the interpersonal aspects of providing effective leadership provide the inspiration for this book and the many real-world examples it contains. They are authors of *Managing People (Including Yourself) for Project Success* (Van Nostrand Reinhold) as well as of numerous papers and presentations on leadership and management.

Joe Stephenson, a graphic designer in Leeds, England, created the cover art for this book.

ONE

What Does It Take to Win?

Identifying the Triangle of Needs and Twelve Key Leadership Characteristics

The Triangle of Needs

The energy of a dog racing team is released when its leader—the lead dog—takes over from a constraining manager and the team races ahead, motivated by the enthusiasm and energy of the lead dog. When project teams are led, rather than merely managed, there is a great release of energy and enthusiasm. Not only can you excite and energize the project team you lead, but you also will, by advancing from being just a manager to being a strong leader, have a more fulfilling and rewarding experience. After all, the lead dog definitely has the best view!

Once a dogsled is up to speed, the amount of resistance that the lead dog and each team member must overcome is small. In order for a project team to quickly get up to speed and encounter a minimum of resistance, an effective project leader sees that the basic needs of the project team members are met before starting work on project tasks. These basic needs fall into three categories:

- ► *Content Needs.* Project scope, budget, expenditures, resources, and schedule.
- ► *Procedural Needs.* How progress is monitored and reported, how issues are resolved, how the team gets paid for doing the project, and how changes in project scope are handled.
- ► *Relationship Needs.* Perceptions of trust, commitment, communication, fairness, respect, participation, and caring.

These essential and equal needs can be represented by the Triangle of Needs shown on page 2, which is based on a mediation model (Moore, 1986). The three sides of the triangle are equal in length because all three needs are equally important in leading a project team to success.

Triangle of Needs.

Both management and leadership functions are essential. Management without leadership results in an uninspired team. Leadership without management can lead to chaos. The difference between those who are only *project managers* and those who are also *project leaders* manifests in the emphasis they give to each side of the Triangle of Needs. Typically, we see project managers spending about 75% of their time and energy focusing on budgets, expenditures, schedules, and scope—all items on the *Content* side of the triangle. Project managers typically spend about 20% to 25% of their time and energy on the *Procedural* side of the triangle. Although 20% to 25% is a significant amount of time, it is often not spent effectively. For example, many project managers don't worry about how to handle changes in project scope until a change actually occurs. The need for a major scope change can upset the customer (the end user of the project team's efforts). A time of emotional upset is not the best time to rationally address how to handle a scope change. It is much more effective to decide how to handle changes in project scope at the beginning of the project. A project leader is forward looking and realizes that spending more time on the *Procedural* side of the triangle at the start of the project will pay big dividends later.

Figure 1-1 starts on page 25

Most project managers seem to spend 0% to 5% of their efforts on the *Relationship* side of the triangle—until something goes wrong on the project, that is. Then, they suddenly must devote 100% of their time to *Relationship* issues to deal with an upset customer or with disgruntled team members. In contrast, project leaders are very aware of the importance of relationship building and, from the start of the project, devote equal attention to the *Relationship* side of the Triangle of Needs. Figure 1-1 identifies some of the activities that must occur on each side of the triangle in various phases of a project. Figure 1-1 can be used as a checklist at the start of each project phase to be sure that a balanced project triangle is established at the start of the project or project phase.

Project leaders know that most project failures occur because of relationship issues rather than lack of a technical skill or lack of knowledge of how to prepare a critical path schedule. This observation is confirmed by research by the Center for Creative Leadership, which surveyed thousands of project participants over three decades to determine the factors that caused projects to succeed or fail (McCauley, 1998). The table below summarizes their findings. Technical ability was a leading cause of project success in only 6% of the projects, and the lack of technical ability was a leading cause of failure in only 10% of the projects. The top cause of success (and failure, when lacking) was relationship-building behavior.

Project Leader Behavior and Project Success/Failure

Behavior	Percent related to project success	Percent related to project problems/ failures*
Relationship building Caring, fairness, demonstrating trustworthiness and understanding	23.8%	45.3%
Cognitive capacity Managing ambiguity, creativity, managing diversity, and system complexity	23.8%	10.5%
Communication Supportive, informing, confronting, presentations, writing	16.4%	5.2%
Self-management Courage, perseverance, self-awareness, time management	16.4%	26.3%
Decisiveness Action orientation, command skills, organizing, prioritizing, results orientation	13.4%	2%
Technical ability Functional skills, specific business knowledge	5.9%	10.5%

* Due to failure to demonstrate the behavior.

Factors on the *Relationship* side of the triangle that describe a high-performing team include the following:

- ▶ Mutual respect and cooperation
- ▶ Clear and positive communication

 ▶ Regular feedback about performance
 ▶ Feeling of appreciation for contribution

The following obstacles to effective teamwork result from inadequate attention to the *Relationship* side of the triangle:

 ▶ Misunderstandings and lack of communication
 ▶ Lack of respect and inconsistent feedback
 ▶ Feeling ignored, unappreciated, and unsupported for efforts
 ▶ Management confusion about team objectives
 ▶ Conflict among team members

The obstacles result from failing to address issues on the *Relationship* side of the triangle. Effective project leaders know that the key to project success is to spend as much time on the *Relationship* side of the triangle as on the *Content* and *Procedural* sides.

The following checklist of characteristics of project leaders demonstrates their attention to the *Relationship* side of the triangle (Culp and Smith, 1992). As you read the list, think about how your project team would rate you on these characteristics. Effective project leaders

 ☐ Are honest
 ☐ Communicate their vision to others, appealing to a common purpose
 ☐ Have energy and enthusiasm springing from a strong belief in purpose
 ☐ Have a positive, can-do attitude
 ☐ Don't control others, but rather trust them and enable them to act
 ☐ Believe in people's potential
 ☐ Are in close touch with those they lead
 ☐ Care about others
 ☐ Understand individual differences and capitalize on these differences to strengthen the team
 ☐ Get everyone involved
 ☐ Respect the aspirations of others
 ☐ Behave as they want others to behave—they walk like they talk
 ☐ Are active, not passive
 ☐ Experiment and take risks
 ☐ Consider mistakes as learning opportunities
 ☐ Welcome change
 ☐ Are good listeners, are open to the ideas of others, and encourage contrary opinions
 ☐ Achieve a team result beyond the sum of individual contributions
 ☐ Recognize and celebrate the contributions of others
 ☐ Proactively consider potential consequences and reactions of others to future actions

Managers—not leaders—honor stability, control through systems and procedures, maintain the status quo, and devote nearly all of their energy to the *Content and Procedural* sides of the triangle. Ineffective managers often attempt to manage their projects from their computer screens, where they view cost and schedule reports and have little interaction with the team—ignoring the *Relationship* side of the triangle. Project leaders certainly pay attention to the *Content* and *Procedural* sides of the Triangle of Needs, but they devote equal energy to the *Relationship* side to bring out the best in people and guide people to success. Managers may buy the *efforts* of the project team members, but they cannot buy their team members' *spirits* and their *minds*. Leaders engage the spirits and the minds of team members so that they want to work toward a shared goal and so that they think differently about their relationship to the team. It is easy to pay someone to do a particular project task, but a leader knows that there will be a much better outcome for the project and the individual team members if they see the importance of the work in relation to both the project and their own individual goals for personal growth. When the team members see how their work fits into the big picture, they are more likely to make contributions to the work that go far beyond a day's work for a day's pay.

Management functions of planning, controlling, and administering are different from the leadership functions of influencing values, motives, and vision. A leader may do both and the functions may often seem to run together, but leadership clearly differs from management. Leaders certainly want the project tasks to be accomplished, but they also want the work to be fulfilling, rewarding, and satisfying to the team members.

There are many books and articles that focus on the *Content* and *Procedural* sides of the Triangle of Needs from the perspective of project management. While recognizing that these two sides of the triangle are important, this book focuses on the *Relationship* side of the triangle to enable managers to become leaders. The need for this focus is well summarized in *The Leadership Challenge* (Kouzes and Posner, 2002):

> Whatever the time, whatever the circumstances, *leadership is a relationship*. Whether it's one-to-one or one-to-many, business as usual or challenges in extraordinary times, leadership is a relationship between those who aspire to lead and those who choose to follow.

The following pages elaborate on twelve key characteristics of effective project leaders that are essential to establishing relationships between the leader and the team.

Twelve Key Leadership Characteristics

Figure 1-2 starts on page 27
As you read about each characteristic, honestly rate where your current level of development of that leadership characteristic falls along the numeric scale presented at the end of each section. Your cumulative rating on these twelve characteristics will give you a starting point in assessing your leadership skills. You can then use the Leadership Assessment in Figure 1-2 as a more detailed

self-assessment and to solicit input from others on your leadership skills. Understanding how others perceive your leadership skills is an important step in deciding where to focus your efforts to become a more effective leader. Effective leaders have a high degree of self-awareness.

Be Honest, Establish Trust

A poll of several thousand people found that honesty topped a survey of characteristics they wanted to see in their leaders (Kouzes and Posner, 2002). Honesty is a key element in satisfying the *Relationship* side of the triangle. The value that people place on honesty is reflected by the characterization of two of the most revered U.S. Presidents. The story of George Washington's confession to his father that he cut down his father's favorite cherry tree—"I cannot tell a lie"—is widely told to illustrate honesty. Abraham Lincoln's nickname of "Honest Abe" was based on the years he spent repaying debts related to a failed retail business early in his career. To demonstrate honesty, you must

Do what you say you are going to do.

Your credibility can be damaged by something as simple as not showing up on time for a meeting you promised to attend or promising to prepare a report by a certain date and missing the deadline.

People watch your feet, not your lips. People trust others who follow through on commitments. Consistency between words and actions is vital. If you as a leader say one thing and do another, your honesty rating plummets. You gain trust by making your values, ethics, and standards clear to your team. You are respected when you live the values you espouse. You are also trusted when you are open and honest about your own mistakes and limitations. Your credibility suffers if you claim to be infallible. Admit your mistakes, make appropriate corrections, and move on.

To establish the *Relationship* side of the triangle for project success, trust is essential. Leaders do not ask others to do something they would not do themselves. Trusting another's competence and judgment results in a greater willingness to be open with that person (Rogers, 1980). A leader establishes a project environment where team members trust that the environment is safe for exchanging ideas and offering differing points of view. Projects succeed when team members and the project leader are open with one another.

To demonstrate honesty and establish trust, you must know your values and live them. Your team needs to know what you stand for. Team members infer your values from their observations of your behavior. As shown in the sketch on page 7, what they see is your behavior, which is made up of your physical, mental, and emotional actions.

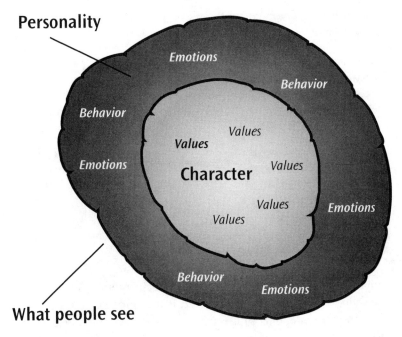

Personality

What people see

Your behavior.

Your values are the relative weight that you place on various principles. Sound principles are those that have stood the test of time, such as honesty, integrity, excellence, compassion, fairness, and trust. A person's mixture of values establishes his or her character and distinguishes one person from another. Defining your values lets your team know what they are supposed to do. Demonstrating honesty and establishing trust requires both values based on sound principles and appropriate behavior.

> ***Both values and behavior are essential;***
> ***neither alone is adequate.***

Your team will give much more weight to the values demonstrated by your behavior as opposed to those that you only speak about. If you preach the merits of high-quality work, your credibility will go up if you are actively involved in quality management training and in performing quality reviews. Also, your values are reflected in the questions you ask. If you ask only about deadlines and budgets, your team will conclude you're not really concerned about quality. If you were, they would expect you to ask about how your customer is feeling about the project. They also watch what you reward. If you say you encourage innovation and risks, then promote and reward those that take reasoned risks, even if they fail occasionally. If a failure related to a reasoned risk is punished, your credibility is destroyed. If you don't set out clear expectations, your team will be second-guessing you. They will not be able to meet your expectations unless they know them. If they see inconsistencies between your stated values and your behavior, they will become frustrated and discouraged.

Now rate where you currently fall on the continuum between the two choices shown below.

| I sometimes adjust my message and actions depending on what others want to hear. | 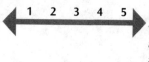 | I consistently express my values in my communication and in my actions. I will take a principled stand even if it is unpopular. |

| I make commitments hoping I can meet them, but sometimes I find that I have promised more than I can deliver. |  | I think carefully before making commitments and I *always* do what I say I am going to do. |

🔑 Get People Involved

Involvement results in commitment. Team members are most energized and productive when they are involved in the project in a significant way. When the project leader encourages collaboration, team members' level of satisfaction and commitment increases. Leaders are catalysts for growth and development.

Start the process by getting the entire team involved in developing a statement of the project vision and a project work plan. During the project, address the *Procedural* side of the Triangle of Needs by establishing procedures so that the team has structured opportunities, such as periodic team meetings, to address project issues. By establishing the structure for input, you are not relying only on spontaneous suggestions and input, which may or may not occur under the pressure of getting the project done. Show that you value input from all team members by talking to individuals between team meetings to solicit their input. Leaders encourage the team to share information, listen to one another, exchange resources, and respond to one another's needs. Leaders set an example of sharing information by being open about the budget available to do the work. We have heard some ineffective project managers say they don't tell team members how much budget they have to do their work because the team will then spend it all. Some project managers give the team an artificially low budget amount, thinking that this will increase the chance of the project being finished within budget. Both of these approaches demonstrate a lack of trust and lack of honesty that is soon sensed by the team. Get the team involved in determining the budget needed to do their tasks and share the reality of what funds are truly available. Work with them to make any needed adjustments in their approach. When team members are involved in establishing their budgets and schedules, they are much more invested in meeting them—and you are doing some valuable work on the *Relationship* side of the triangle by demonstrating your trust in their abilities.

Allow and encourage individuals to make decisions. If you horde information and make all of the decisions, no one will be prepared to step in if you are absent at the time when an emergency arises and a critical decision needs to be made. Getting others involved creates synergy between the *Content, Procedural,* and *Relationship* sides of the triangle, but adequate attention to the *Relationship* side is critical for the synergy to occur.

Leaders keep no secrets about the project goals, status, problems, and successes, and leaders clearly define each person's role in the overall picture. They keep the team informed through periodic meetings and individual discussions so that the team members know what is needed to achieve successful results. The involvement of everyone, in an environment of openness, fairness, respect, and honesty, is a key to project success.

Now rate where you currently fall on the continuum between the two choices shown below.

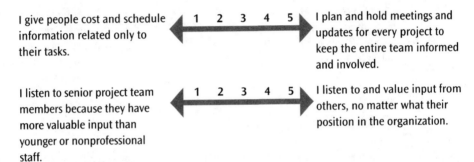

I give people cost and schedule information related only to their tasks.	1 2 3 4 5	I plan and hold meetings and updates for every project to keep the entire team informed and involved.
I listen to senior project team members because they have more valuable input than younger or nonprofessional staff.	1 2 3 4 5	I listen to and value input from others, no matter what their position in the organization.

Encourage Contrary Opinion

A team of similar-thinking people or people with similar personal styles (see Chapter Two) can develop blind spots, resulting in "group think." The results can be disastrous. The Space Shuttle *Challenger* disaster and the Hubble Space Telescope problems have been linked to decision processes at NASA, where group-think symptoms surfaced (Robbins, 2002).

The free expression of differing opinions is good for the health of the project and the morale of team members. A leader recognizes and truly *values* the diversity of the team members' styles, talents, and opinions as a resource, not a liability. Leaders seek out contrary opinions and carefully consider them. Leaders encourage differences and discourage indifference. You may want to appoint one team member to play the role of devil's advocate by openly challenging the majority position and offering a differing perspective. We have seen some teams use a technique in which team members wear different-colored hats that represent their various roles during a project meeting. For example, team members wearing white hats say only positive things, those wearing black hats act as the devil's advocate, those wearing green hats come up with solutions to the arguments of the devil's advocates, and those wearing blue hats can only ask questions of the others. At some point in the exercise, the members switch hats so that they are forced to see other points of view.

Focus on the *Relationship* side of the Triangle of Needs to maintain a safe environment for team members to respond, and then give careful consideration to their opinions. Another approach is to have the team members talk about the risks involved in a decision before talking about the advantages of the decision. As a result, the team is less likely to stifle differing opinions and more likely to make an objective decision.

Now rate where you currently fall on the continuum between the two choices shown below.

"If it ain't broke, don't fix it." It would be a waste of my time to even talk about it.	1 2 3 4 5	I enjoy and value diverse opinions. I find them a good way to build consensus and reach the best decisions.
When I encounter an opinion different from mine, I typically offer an argument to support my position.	1 2 3 4 5	I seek out those who usually have different ideas from mine, solicit their input, and carefully consider their ideas before responding.

Establish a Vision

A leader develops and consistently communicates a vision of what the completed project will look like and how each of the project tasks will contribute to achieving that vision. In doing so, the leader addresses the *Content* and *Procedural* sides of the Triangle of Needs. To address the *Relationship* side of the triangle, leaders make sure that their vision comes alive and stays alive by involving the team in developing a statement of the vision and project objectives, continually communicating the vision and these objectives, and keeping in close contact with the team. Leaders stress the aspects of the vision that address what the team members' value, strengthening the sense of common purpose. They get out of their offices to spend time with the team, to learn what the team members want from the project.

Visions motivate and are challenging, but they are not so far beyond existing capabilities as to be hopeless. They offer a chance to be tested and are broad enough to allow teams to use their own ingenuity in solving problems. They provide a chance to develop and learn. They look to the future and provide a chance to make a difference. The vision gives everyone on the team a chance to do something that clearly contributes to the end result. An effective vision attracts people to it by the force of its appeal.

At a time when network news programs were 15 minutes long, Ted Turner had a vision of a 24-hour news network. Most thought him crazy; few gave him a chance of success. Yet his belief in his vision and his leadership created the Cable News Network. Fred Smith was told that his vision of a service that would deliver a package anywhere overnight was impractical, but Fred followed his vision and created Federal Express. Bill Gates has made his vision very visible to those at Microsoft by installing a plaque on the Microsoft campus in Redmond, Washington, that reads, "Every time a product ships, it takes us one step closer to the vision: a computer on every desk and in every home."

Project leaders deliver their messages with enthusiasm and highly visible energy. Their positive attitude conveys warmth and friendship. They use language that paints a vivid picture of their vision in a positive, hopeful way. Martin Luther King's "I have a dream" speech exemplifies many of the features of an effective vision because it

- ▶ Appealed to the common purpose of his audience
- ▶ Used examples that all could relate to
- ▶ Painted vivid pictures of a positive future with words
- ▶ Was positive and hopeful
- ▶ Was passionately delivered

The speech is brief, only about 10 minutes long. Rent or buy a video of King's speeches. Watch it to experience an effective vision—and look for techniques he used that you can also use.

Now rate where you currently fall on the continuum between the two choices shown below.

I limit the information I give to an individual to the tasks he or she is assigned so he or she won't be distracted with information about tasks being done by others.	1 2 3 4 5	I involve the team in establishing project vision, objectives, and approach. I make sure that all team members understand how their tasks relate to other tasks and to achievement of the project vision.
Once a task is assigned, I leave the person alone until he or she decides to report back to me.	1 2 3 4 5	I frequently seek out team members to discuss the project objectives and how their ongoing work is advancing the team toward those objectives.

Take Risks

Leaders take reasoned risks and encourage others to do so. Leaders recognize that any project that is going to make a difference will involve risk. A leader makes it safe for team members to fail and learn from mistakes. Baseball legend Babe Ruth, best known for successfully hitting 714 home runs, also struck out 1,330 times. Babe lived the advice he offered others: "Never let the fear of striking out get in your way."

Leaders persist even when risks lead to "failures." Abraham Lincoln's first business failed in 1831. He turned to politics and was defeated for the Illinois State

Legislature in 1833. His sweetheart died in 1835. He had a nervous breakdown in 1836. He ran for Congress in 1843 and was defeated. After being elected to Congress in 1846, he lost his seat in 1848. He was defeated for the Senate in 1855. He lost his bid for vice president in 1856. He was defeated again for the Senate in 1858. Today, he is recognized as a great president, despite his many "failures." Theodor S. Geisel wrote a children's book that was rejected by twenty-three publishers. He persisted, and the twenty-fourth publisher sold six million copies of his first "Dr. Seuss" book. R.H. Macy failed in retailing seven times before his store in New York succeeded. Thomas Edison said he failed his way to success because he tested thousands of ideas for filaments for electric lightbulbs that failed before he found one that worked.

Project leaders encourage team members to keep looking for a better way to do things and accept that innovation is likely to involve some "failures." The leader focuses on what can be learned from the failure. Consider the story of the beginning skier who proudly skied up to his instructor the day after his first lesson and announced he had a great day because he did not fall once. The instructor's response was not what the skier expected: "If you didn't fall, you learned nothing today. Your skiing will be no better tomorrow than it was today."

When a failure occurs, the words you use and the way you deliver them will affect people's willingness to take risks in the future. Separate the performance from the performer. Avoid saying things like, "This certainly did not turn out the way I wanted." Even if the results weren't what was envisioned, you can say, "You carried out the plan we agreed upon. The thorough job you did sets a great example for the team. Let's talk about what we all learned from this and what we will do next." Focus on what has been learned, not on the problems. Again, spending time and energy on the *Relationship* side of the Triangle of Needs will likely have benefits that carry over to the *Procedural* and *Content* sides. Project leaders work with their teams to identify what could be done better next time. Record the notes and make them available to the team. One company we know keeps a notebook of lessons learned from projects and provides a copy to every project leader. It is acceptable to make a new error, but it is not acceptable to repeat one that is already documented in the notebook. New errors are debriefed, documented, and added to the notebook. Review past lessons learned before you start your next project.

Sometimes the risks that need to be taken fall on the *Relationship* side of the triangle. We have been involved in projects where some of the participants did not trust others that were going to be involved in the project based on their past relationships. Although it may appear risky to attempt to resolve long-standing relationship issues at the start of a project, a leader is not deterred from doing so. President Jimmy Carter faced a substantial risk of failure when he asked the leaders of Israel and Egypt to sit down together at Camp David in an attempt to resolve long-standing, bitter differences. Although it was a long process—no doubt with periods of discomfort for all—Israel and Egypt reached an agreement and the Camp David accord was signed. An effective leader gets distrustful parties together and works with them to break through relationship barriers that otherwise may prevent a successful project.

Now rate where you currently fall on the continuum between the two choices shown below.

If a task is not going well, I remove the assigned person from the task and either finish the work myself or assign it to someone else.	1 2 3 4 5 ⟵⟶	If a task is not going well, I meet with the assigned individual to provide support and direction and use the situation as a learning opportunity.
I do not encourage the team to take risks or change an existing practice because I am more concerned about the negative results that may occur than the potential benefits.	1 2 3 4 5 ⟵⟶	I encourage the team to take reasoned risks by recognizing the good tries as well as the successes.

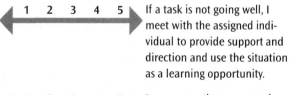 Create a Positive Environment

In a positive project environment, team members enjoy their work because they are growing both personally and professionally. They have a clear vision of where they're headed and the knowledge that they will help shape their own futures. A team that enjoys what they are doing is more energetic and involved than a team doing boring, routine tasks in a somber atmosphere. Certainly some routines are important, such as seeing that quality control procedures are routinely applied. The question becomes, Which routines are critical to project success and which are not? When a routine begins to stifle creative thinking or create roadblocks to progress, the leader gets rid of or changes the routine. Wal-Mart Stores has a company philosophy of ridding their organization of needless routines. They have called it their ETDT (an acronym for "Eliminate the Dumb Thing") program. How many "dumb things" come to mind in your organization or in the procedures being used on your projects?

We once had an office located in a building where Microsoft had some offices. In the few glimpses we got through their opening and closing doors, we could often see colorful balloons floating around the work spaces and boisterous, barefooted staff walking around in shorts or riding skateboards in the halls. Although they worked early and late hours, they always looked like they were having fun.

The passion with which the leader works toward the project mission excites others. You may have heard that if you want others to smile, smile at them. Similarly, if you want your team members to be passionate about the project, show them your passion for the project. A positive environment offers team members challenges, opportunities to do different things and explore new areas. A leader demonstrates a sense of humor. Herb Keller's lighthearted and humorous approach has often been cited as a key reason for the success of Southwest Airlines and is reflected by the encouragement of the airline's flight attendants to sing humorous songs and tell jokes to the passengers.

Now rate where you currently fall on the continuum between the two choices shown below.

This is a serious business and there is no place for idle chatting or frivolity.	1 2 3 4 5	Life is too short. Let's enjoy what we are doing. I look for fun ways to celebrate our successes.
I make sure the team knows and follows established procedures because they provide the best way to get the project out the door.	1 2 3 4 5	Our work is exciting and compelling, and it provides an opportunity for me to inspire others to find new ways to reach our project goals.

Challenge Limiting Beliefs

For many years, it was believed that running a mile in 4 minutes or high-jumping 7 feet was physically impossible. Now these "impossible" goals are routinely exceeded. Once the mental barrier was broken, many others began to run a mile in less than 4 minutes and to high-jump more than 7 feet. Project leaders recognize that some team members' performance may be limited by mental, self-imposed barriers and will look for ways to help team members break through these barriers.

Project leaders find opportunities for people to explore new ground on their projects. They work with each team member to see what personal growth objectives might be achieved through appropriate project assignments that will help team members break through their own self-imposed, limiting beliefs about their abilities.

As the project proceeds, a project leader will ask the project team what needs to be changed. The leader may ask, "What is the basis for that assumption? Where does that premise come from? What other approach might work here?" The leader devotes part of every team meeting to brainstorming innovative approaches. The leader gives everyone a chance to participate in shattering a limiting belief by coming up with a new approach, procedure, or process. In doing so, the project leader is addressing the *Relationship* side of the Triangle of Needs, and the results may well affect the *Procedural* and *Content* sides of the triangle by generating changes in the project approach and scope.

Now rate where you currently fall on the continuum between the two choices shown on page 15.

| I prefer to rely on approaches that have worked for me in the past. | 1 2 3 4 5 | I am energized by finding new ways to achieve challenging goals, and I encourage the team to seek new approaches. |
| I tend to rely on historical experiences with others to predict future interactions. | 1 2 3 4 5 | I approach each new situation with others as an opportunity to further all of our goals. |

Choose Your Reactions

Effective project leaders take the initiative and do whatever is necessary, and proper, to get the job done, often without waiting for perfect information. They choose to treat decisions that others find stressful as an opportunity to create positive results. They make decisions based on the best available information and move on. People follow those who take decisive, reasonable action.

Effective leaders own their behavior. Chapter Two is devoted to understanding personal behavioral style preferences. Effective leaders know that their behavior is a result of their own conscious choice rather than a result of an uncontrollable reaction to external circumstances. Your behavior is a function of your decisions, not outside conditions. No matter what happens, effective leaders know they can choose how they react.

How you REACT to the issue IS the ISSUE.

You may have, at some time, said something like, "John isn't going to finish his task assignment on time, and it looks like he is going over budget again. I'm upset because this has happened to me too many times." Being "upset" is a mental, emotional response that isn't related to the physical world. It is a response that you are choosing. You can choose to react differently and in a more positive way by recognizing that it's not the event that causes you to be upset, but your reaction to the event that causes your response. "I am upset" is the only important part of your statement. Ask yourself, "Why am I upset?" Are you upset because this is a continuing pattern for John and you have failed to provide the support or direction needed to resolve it? Other factors might have been involved: you had to change a flat tire in the rain on the way to work; you had a call from an unhappy customer; you shouted at your kids this morning and feel bad about it; the person you're upset with reminds you of some other bad experience. We tend to react out of habit based on conclusions drawn from our experience. We sometimes project these experiences onto other people and expect the worst or overreact. Being aware that it is an internal process of yours that is causing the upset, *not* the external event, is a big step toward choosing your reaction. Next, choose to react differently based on the individual situations. Rather than being upset in this example, you could choose to meet with John and explore what support, assistance, or direction is needed to break the pattern of missed deadlines and blown budgets.

We have seen instances where a project manager is upset because of comments made about the project during a project review or value engineering analysis. The upset may surface as a defensive attitude, in which the manager takes the comments as a personal attack and dismisses the value of the comments. Although the project manager may be blaming the upset reaction on the external actions of others, the cause of the upset is an internal one that is possibly related to a lack of confidence, a self-questioning of abilities, or anxiety over appearing weak by accepting the ideas of others. The project manager could choose to react in a more positive way by objectively considering the content of the comments, adopting those of value, and moving on with the project work. In choosing to do so, the project manager is focusing on the content of the message being conveyed by the ideas of others and allowing any emotional energy attached to that message to pass by without affecting him/her.

An effective leader breaks the potentially negative connection between outside events and his/her responses by choosing to respond in a constructive way.

Now rate where you currently fall on the continuum between the two choices shown below.

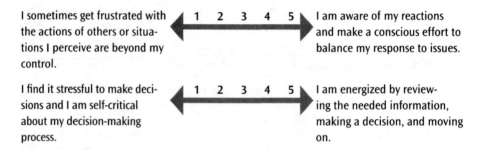

I sometimes get frustrated with the actions of others or situations I perceive are beyond my control.	1 2 3 4 5	I am aware of my reactions and make a conscious effort to balance my response to issues.
I find it stressful to make decisions and I am self-critical about my decision-making process.	1 2 3 4 5	I am energized by reviewing the needed information, making a decision, and moving on.

Recognize Performers

Have you ever heard your team members say, "We are just sick and tired of all the praise we are getting"? We would be very surprised if you have. Yet, to be recognized and appreciated is a basic need on the *Relationship* side of the triangle. An example is offered by a theme park that passed out blue cards known as Warm Fuzzy Awards for good performance (Zemke and Anderson, 1997). The date and reason for the award was jotted down on the back of the card and the cards could be redeemed for prizes. The only problem encountered with the program was that the employees did not exchange their cards for prizes. An employee focus group said that the psychological value of the award cards outweighed the value of the prizes. One employee said, "When I'm having a bad day, I take out my stack of Warm Fuzzies and reread the notes on the backs. The nice things people said about me make me feel better. That's more important than any prize I could buy for turning in the cards."

Recognition is a powerful tool when it is sincere, is offered immediately, and specifically describes what constituted the outstanding performance. It is important to be specific. Don't limit your praise to "good job" or "keep up the good work." It is much more effective to say, "Jane, I really appreciate the data analysis you completed prior to the meeting today. It allowed us all to decide on our next steps and saved us a lot of time. Thank you for the effort you put in to do it so quickly." Don't wait for formal performance reviews to recognize performers—do it at the time the performance occurs.

Rewards need to address the needs of the person you're recognizing. What's rewarding for one person may not be rewarding for someone else. In one case, the president of a large company thought taking a staff member to lunch would be a reward. One staff member who had not met the company president was terrified by the thought of spending an hour alone with the president. He became too ill to go to lunch.

Now rate where you currently fall on the continuum between the two choices shown below.

Besides the common courtesy of saying "thank you" for tasks done, I rely on our formal annual performance review system to provide feedback to team members.	1 2 3 4 5	I provide specific, detailed feedback and visible recognition immediately after an accomplishment by an individual or the team.
I delay addressing project performance problems because sometimes they seem to take care of themselves.	1 2 3 4 5	When there is a project performance problem, I promptly address it by offering constructive feedback.

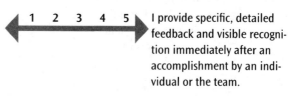

Be Decisive and Competent

A team will follow someone they believe knows what he or she is doing. Competence in this case does not necessarily mean that the leader has to have the most outstanding technical skills on the team. The Manhattan Project, which produced the atomic bomb that ended World War II, was not led by an atomic scientist but rather by General Leslie Groves, someone who did not understand atomic physics. Competence involves getting things done for your project team that must get done and seeing they have the information and tools needed to succeed. Technical credibility and related knowledge of the *Content* and *Procedural* sides of the Triangle of Needs may be necessary, but that, in itself, is not enough. On the *Relationship* side of the triangle, the competent project leader demonstrates the ability to effectively plan, encourage, coach, and inspire the team. Over time, the leader's track record of having successful teams becomes a demonstration of competence.

Project leaders have a never-ending desire to learn and grow, ever increasing their areas of personal competence. They continue to set and achieve personal improvement goals throughout their careers. Project managers often find that their interpersonal skills must be improved to make the transition to project leader. They recognize that most project problems arise from relationship expectations that are not met, rather than from technical issues. So, successful project leaders are very self-aware of the impacts of their behavior on others, work to improve their relationship skills, and pay attention to the human side of their projects, that is, the *Relationship* side of the Triangle of Needs.

The education and training experience of most project managers emphasizes the technical aspects of projects, not people skills. Participating in courses and seminars to improve your interpersonal skills demonstrates to others that you are serious about improving the team atmosphere and becoming a leader.

Now rate where you currently fall on the continuum between the two choices shown below.

It is important to me that others perceive I have the most knowledge about the project. I am sometimes uncomfortable with team members who have more experience.	1 2 3 4 5	I am confident in my abilities and I draw upon team members in my areas of weakness and in their areas of strength.
Setting personal growth goals is something for our younger team members to do, not for an experienced person like me.	1 2 3 4 5	I continually set personal growth goals and achieve them.

Align Individual and Project Goals

When giving adequate attention to the *Relationship* side of the Triangle of Needs, leaders align the project goals with the goals of the individuals on the team. This alignment occurs when individuals feel that their contributions to the project also are contributing to their personal goals and growth. Certainly, there will be project situations where budget or schedule constraints make it difficult to have everyone's tasks align with their personal growth goals. However, look for every opportunity to do so. At the start of the project, discuss a project learning plan with each team member. Develop the plan by asking if there is a task that he or she would like to work on to achieve a personal growth goal. In the process, learn all you can about each team member's growth goals so that you can be alert for opportunities on other projects. Don't assume that budget constraints will preclude such an opportunity. In one case, one of our team members wanted to learn more about the public involvement aspects of the project approval process. She was so eager to do so that she volunteered to attend the evening public hearings on her own time to assist by taking notes and interacting with the public before and after the formal presentation.

A superior coach of an athletic team comes to know all players well enough that their team assignments best match their skills *and* desires. The individual wants the team to win because the team provides the vehicle for achieving his/her own goals. Amazing results can occur when team and individual missions are in alignment because both the minds and hearts of the team members are drawn into action. Team results can exceed the sum of the individuals' abilities. The Boston Celtics basketball team won sixteen titles without ever having the league's individual leading scorer.

By focusing on the *Relationship* side of the Triangle of Needs, a project leader spends time with the team to listen to them and understand their goals and needs. Determine what part of the project most appeals to each team member and, whenever possible, assign him or her to that part of the project. If a team member has a goal of professional recognition, give that person the assignment of preparing and presenting published papers on the project. If one team member's goal is to get experience in all phases of a project, give that person the assignment that he or she has not yet had. Team members' energy for the assignment involving new work that they want will be much greater than if they are given an assignment involving work that they have done many times before.

Now rate where you currently fall on the continuum between the two choices shown below.

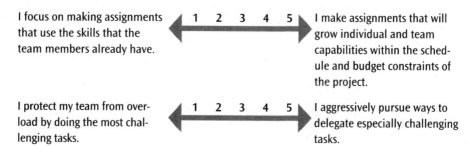

I focus on making assignments that use the skills that the team members already have.   1   2   3   4   5   I make assignments that will grow individual and team capabilities within the schedule and budget constraints of the project.

I protect my team from overload by doing the most challenging tasks.   1   2   3   4   5   I aggressively pursue ways to delegate especially challenging tasks.

Establish and Achieve Doable Goals

The successful professional football quarterback Fran Tarkenton said, "Beware the big play; the 80-yard drive is better than the 80-yard pass." Similarly, project leaders get people to tackle big challenges by breaking them down into smaller, doable tasks—lots of 8-yard plays rather than the one big 80-yard play. Successful completion of each small task sets forces in motion that encourage steps toward success on another task. Your team's confidence level will rise as their desire to succeed is reinforced. The small wins deter opposition. After all, it is hard to argue with success. Resistance to later proposals drops. Your team will see that you are asking them to do things that they know they are capable of doing.

To be doable, each task has its own description, schedule, budget, and personnel assignments. As one task builds upon another, the overall project moves to completion. Trying to accomplish too much at once or presenting an overwhelming task in one play doesn't work. Team confidence grows as each 8-yard pass is completed; compare that to the morale of a team that consistently has three incomplete 80-yard passes and must then punt the ball away to the opponent. Proper attention to the *Content* side of the triangle—by establishing doable tasks—can help the *Relationship* side by establishing achievable goals that build team confidence.

It is critical that you give your team constant feedback as each task progresses. They must know how they are doing before they tackle the next task. You wouldn't expect a football team to be able to adjust their strategy if they didn't know the score until the game was over.

Now rate where you currently fall on the continuum between the two choices shown below.

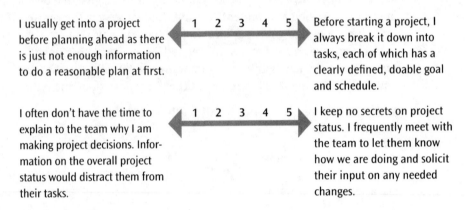

I usually get into a project before planning ahead as there is just not enough information to do a reasonable plan at first. | 1 2 3 4 5 | Before starting a project, I always break it down into tasks, each of which has a clearly defined, doable goal and schedule.

I often don't have the time to explain to the team why I am making project decisions. Information on the overall project status would distract them from their tasks. | 1 2 3 4 5 | I keep no secrets on project status. I frequently meet with the team to let them know how we are doing and solicit their input on any needed changes.

Fire Fighter or Fire Lighter?

> *Managers fight fires . . . leaders light fires.*
> —Perry Pascarella and Mark Frohman

This quotation refers to the time that managers spend acting as controllers and administrators by fighting destructive fires caused by inadequate leadership. On the other hand, leaders are lighting motivating fires that inspire others to new levels of performance.

The preceding pages provided two five-point scales for each of twelve leadership characteristics so that the maximum total points for the positive leadership characteristics at the five-point end of the scale is 120. If your total is 100 to 120, you probably already have ignited some large, roaring, motivating, and inspiring fires for your team. If your total is 75 to 100, you probably have ignited a scattered fire or two, but there is a lot of room to build more motivating fires. If your total is less than 75, you need to start rubbing some sticks together.

No matter what your total, you can assess your areas of strengths and development by completing the Leadership Assessment (Figure 1-2) presented in Exercise 3 at the end of this chapter. This is a self- and 360-degree–assessment tool developed and widely used by Smith Culp Consulting. Ask others to complete the Leadership Assessment for you to obtain a more accurate assessment of your current leadership skills.

The Lead Dog Can Save the Day

Leadership skills are an observable set of abilities that can be honed through awareness and practice. With enough practice, they can be integrated into your behavior so that they appear intrinsic and intuitive to others. The intuitive integration of these skills can sometimes provide seemingly amazing guidance, avoiding project disasters. Bob Klasges, a retired dogsled racer, relates a story in *Born to Pull* that demonstrates how the lead dog can save the day (Cary and deMarcken, 1999). He had hooked up a dog team to pull him up to a river to fish with his favorite lead dog, Nanook, leading the team. While fishing, a blizzard blew in, causing a total whiteout. Bob had no idea where the trail was but knew there were some dangerous spots where the river wasn't frozen. It was too risky to wait out the storm. With no other choice, he packed up his fishing tackle, climbed on the sled runners, and said to his lead dog, "Nanook, take me home." They started out into the blinding snowstorm. A short way down river, they came upon two snowmobilers who were parked on the ice. Their trail had drifted over and they had no idea where they were. It was growing dark. Bob explained that he was relying on Nanook to find the way out and that if they wanted to be alive in the morning, they had better follow. Bob took a flashlight out of his packsack, lashed it to shine backward from the sled so the snowmobilers could follow, and yelled "Hike" to Nanook. Nanook led the team through miles of darkness and blinding snow to Bob's home, where his wife was worriedly peering into the night. In moments, they were all within the warm confines of home, drinking hot coffee and laughing about their ordeal. Only Nanook knew how close they had come to disaster in the form of open water. Similarly, an effective project leader can guide the project team through difficult and confusing circumstances to avoid project disasters—and with enough practice, his/her guidance will seem intuitive.

Summary

Every project team has basic needs that fall into three categories: *Content* needs, *Procedural* needs, and *Relationship* needs. These needs can be represented as the three sides of the Triangle of Needs, a triangle that has three equal sides because all three categories require equal attention. Project *managers* spend most of their effort on the *Content* and *Procedural* issues, neglecting *Relationship* issues until something goes wrong on the project. Project *leaders* know that most project failures occur because of *Relationship* issues and, although they deal with *Content* and *Procedural* issues, they spend an equal amount of effort on *Relationship* issues. Project *leaders* demonstrate several key characteristics:

they are honest, get people involved, encourage contrary opinions, establish a vision, take reasoned risks, create a positive environment, challenge limiting beliefs, actively choose their reactions rather than passively reacting, recognize performers, are decisive and competent, align individual and project missions, and establish doable goals.

Exercises

1. Keep a log of how you spend your time for a week by noting the side of the Triangle of Needs involved in each of your activities. What percentage of your time did you spend on each side of the Triangle of Needs? Enter the percentages next to each side of the triangle below. What can you do to get a better balance among the *Content, Procedural,* and *Relationship* sides?

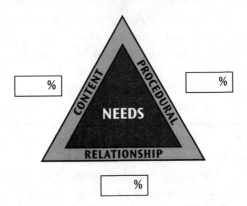

2. Figure 1-1 lists several project leadership and management responsibilities related to each side of the Triangle of Needs. Fill in the blanks with other responsibilities that you think you should meet to be an effective project leader. Use this as a checklist for each project phase.

3. Use the Leadership Assessment shown in Figure 1-2 (or other commercially available leadership skill assessment instruments as described in Campbell, 2002; Nilsen and Campbell, 1998; Kouzes and Posner, 2001) to make a self-assessment of your leadership skills and to get feedback from others. In conjunction with the University of Washington, Smith Culp Consulting has developed an on-line version of the Leadership Assessment tool shown in Figure 1-2 that can be used to ease the input and analysis procedures. Information on use of the on-line version can be found at www.smithculp.com. The Leadership Assessment in Figure 1-2 is designed to be used by others but you can also use it for your self-evaluation. To be an effective leader, you should have an understanding of how your leadership skills are viewed by others. Effective leaders have a good sense of self-awareness. A very important and effective way to increase your self-awareness is to ask for feedback from others about their perceptions of your leadership skills. The Leadership Assessment form is a tool for doing so.

Figure 1-3 starts on page 29

▶ From your work environment, ask three to five peers, three to five people who report directly to you, and the manager to whom you report to act as observers and provide their perceptions of your skills by anonymously filling out copies of the Leadership Assessment and returning them by a specified date. Compile the ratings using the Rating Summary Form in Figure 1-3. It is important that these observers know that their responses will be confidential and that their input is to be used strictly by you as your individual improvement and development tool and not in any type of performance evaluation.

▶ Choose observers from your work environment in your office location that you work with on a day-to-day basis and interact with frequently throughout the year.

▶ Do your best to get eight to ten observers total, five observers at a minimum.

▶ Avoid choosing observers in different geographic locales (an exception being if your manager is in a different geographic locale) or customers because they interact with you much less frequently than those that you work with on a day-to-day basis.

▶ Personally deliver the Leadership Assessment survey to all observers, and let them know how much you value their honest feedback. Give them a date by which you want the Leadership Assessment returned, and tell them to whom the assessment should be given.

▶ Keep a list of your observers so you'll be able to remind them to submit their completed surveys.

▶ Stay in touch with your observers to make sure they return their surveys by the due date.

▶ Using the Rating Summary Form (Figure 1-3), tabulate the average observer ranking and your self ranking for each item. Smith Culp Consulting's on-line version of the Leadership Assessment tool automatically compiles the results and provides added assurance of confidentiality.

Figure 1-4 starts on page 32

▶ Evaluate the results. Note the instances where the observers' and your self-rankings are similar and different. It is also useful to prepare a graph (see the example in Figure 1-4) that summarizes the percentage of favorable ratings (4 or 5 rating), neutral ratings (3 rating), or unfavorable ratings (1 or 2 rating) for all of the items in each category of characteristics. Such a graph presents a quickly grasped picture of relative strengths and areas for improvement. It is preferable to involve a third-party professional with experience in 360-degree feedback in interpreting the results and providing coaching based on the results. The potential for stress caused by the fact that others may perceive your skills differently than you do can be better addressed with such assistance. If such assistance cannot be made available, there is still much to be gained by carrying out the assessment. When you see the results, then use the worksheet in Figure 1-5 to do the following:

Figure 1-5 starts on page 33

- List up to three of your leadership characteristics that both you and others have perceived as strengths that you tend to rely on, that seem most important to you, and that you exercise comfortably and frequently. Ask yourself what you do that accounts for these high scores. What can you do to further reinforce these skills?
- List up to three of your leadership characteristics you were most surprised to see that others view as strengths that you did not. These are likely to be strengths that you are less comfortable using. More frequent use of these skills may offer the most immediate potential for improved leadership. List ways that you could use these strengths more often.
- Determine which of the leadership skills you were most surprised to see where others' rankings were lower than your own. These are development needs that you have not previously recognized. These needs represent the greatest opportunity for growth.

Figure 1-6 starts on page 34

4. Considering the results of the above exercise, pick the three characteristics you feel are most in need of improvement and write down what actions you are going to take and when you are going to take them. These may be existing strengths that could be even more effectively used or existing development needs. Review the checklist in Figure 1-6 for ideas. Also, review Chapter Two and the checklists at the end of Chapter Two for more ideas. Share your action plan with a peer and ask that person to check in with you on your progress at specific times.

5. An important step is to share the feedback and your action plan with those who offered you feedback. You will set an example by being open about information that is relevant to improving the team, and you will be demonstrating honesty, trust, and teamwork. Talk about your highest scores and ask how you can become even better. Talk about the areas of improvement, ask for examples, and get feedback on how you can improve. Thank them for their feedback. You can do this either in a group setting or in one-on-one discussions, depending on your comfort level and that of the people who gave you the feedback. The fact that you are sharing the feedback will mean a lot to your team. Provide specifics about your action plan to improve, and ask the team to hold you accountable and to give you positive feedback when they see you are doing what you said you would do. Do not under any circumstances be defensive or justify your past actions or challenge their ratings and feedback. If you do so, it will be the guarantee that you will never again get any honest feedback. Rather than asking, "Why did you rate me that way?" ask, "Can you give me an example? How could I better do [that task]?" Listen without reacting, and thank the person for the feedback. Afterward, think about it and develop an action plan. Encourage others to use the Leadership Assessment process to improve their own leadership skills, reminding them that it doesn't matter whether they are in a management position. Everyone functions as a leader at some time.

Figure 1-1 *Checklist of Project Leadership Responsibilities and the Triangle of Needs*

Project phase	Content	Procedural	Relationship
Planning/ scoping	☐ Define project scope to meet objectives	☐ Conduct planning workshop	☐ Define end-user/stakeholder needs and goals
	☐ Determine conceptual feasibility	• Define goals, roles	☐ Involve the team to understand big picture
	☐ Select best option for design	• Determine constraints • Flowchart process	☐ Start building team and trust/communication
	☐ Identify detailed tasks	• Identify potential issues and solutions	☐ Involve decision makers early on
	☐ Identify milestones	☐ Write project work plan	
	☐ Identify systems impacted	☐ Review scoping do's and don'ts	☐ Reach agreement on terms and definitions
	☐ Define training needed		☐ Identify concerns/impacts to team
	☐ Review system database	☐ Define how tasks fit together	
	☐ Identify resources	☐ Define accountability	☐ _____
	☐ _____	☐ Define method of conflict resolution for different interpretations	☐ _____
	☐ _____		☐ _____
	☐ _____		☐ _____
	☐ _____	☐ Define process for resolving schedule conflicts	☐ _____
	☐ _____		☐ _____
	☐ _____	☐ Define customer feedback/ communication plan	☐ _____
	☐ _____		☐ _____
	☐ _____	☐ _____	☐ _____
	☐ _____	☐ _____	
Design/ development	☐ Prepare detailed scope, schedule, and budget	☐ Delegate and assign tasks	☐ Coordinate team and information
		☐ Monitor task progress	
	☐ Design parameters	☐ Hold project review meetings	☐ Get team necessary resources
	☐ Develop specs		
	☐ Ensure quality control	☐ Build consensus	☐ Solve problems and resolve conflict
	☐ Identify available resources	☐ Frame/reframe issues to reach solutions	☐ Understand different working styles
	☐ Develop standards	☐ Monitor schedule/budget	
	☐ Develop operating documentation	☐ Define communication plan	☐ Identify impacts of projects on groups
		☐ Define decision-making process	☐ Solicit feedback at interviews
	☐ _____	☐ Develop contingency plan	
	☐ _____		☐ _____
	☐ _____	☐ _____	☐ _____
	☐ _____	☐ _____	☐ _____
	☐ _____	☐ _____	☐ _____
	☐ _____		

continued on next page

Figure 1-1   *Checklist of Project Leadership Responsibilities and the Triangle of Needs (continued)*

Project phase	Content	Procedural	Relationship
Implementation/start-up/construction	☐ Monitor quality to specs ☐ Finish punch lists ☐ Facilitate debugging ☐ Identify what is missing ☐ Conduct training ☐ Update ongoing issues list ☐ Develop maintenance documentation ☐ _____ ☐ _____ ☐ _____ ☐ _____ ☐ _____ ☐ _____ ☐ _____	☐ Solve problems ☐ Perform value engineering sessions ☐ Troubleshoot issues ☐ Hold postmortem debrief meetings ☐ Handle overlooked items, changes ☐ Define launch process ☐ Conduct status checks ☐ Document issues/changes ☐ Make timeline adjustments on overlooked items ☐ _____ ☐ _____	☐ Monitor end-user/stakeholder satisfaction on initial operations ☐ Hand over from design to start-up ☐ Communicate changes ☐ Communicate launch process ☐ Solicit feedback on training ☐ _____ ☐ _____ ☐ _____ ☐ _____ ☐ _____
Follow-up/ongoing support/operations	☐ Provide technical support of product/facility ☐ Tweak system/modify facility ☐ Establish maintenance plan ☐ Update changes to documentation ☐ _____ ☐ _____ ☐ _____	☐ Review closeout checklist for project ☐ Get feedback to designers on system/facility ☐ Document changes ☐ Define process for handling feedback ☐ _____ ☐ _____ ☐ _____	☐ End user feedback ☐ Hand over to ongoing support/customer care ☐ Sign off on closure ☐ Acknowledge contributions ☐ Party/celebration ☐ Notify everyone of documentation changes ☐ _____ ☐ _____

Figure 1-2 *Leadership Assessment*

Assessment of: *(Individual's Name)*

Observer Instructions. The purpose of this assessment is to provide a comparison of the individual's self-evaluation of leadership characteristics with the evaluations of others (observers). The resulting profile will provide information so the individual can compare his/her self-evaluation with the perceptions of others and identify areas for improvement. Indicate how descriptive the statement is of the behavior of the individual you are rating by circling the appropriate rating number. Rating numbers are defined as follows:

1 = Never; 2 = Seldom; 3 = Sometimes; 4 = Usually; 5 = Always.

Work rapidly. Your first impression is usually most useful.

Leadership characteristic	Rating
Is honest, establishes trust	
• Follows through on all commitments; does what he/she says he/she is going to do	1 2 3 4 5
• Sets a personal example of what he/she expects from others	1 2 3 4 5
• Makes beliefs and values known, and lives them	1 2 3 4 5
• Admits personal mistakes, learns from them, moves to correct the situation	1 2 3 4 5
Gets people involved	
• Is open and approachable; keeps in touch with team members	1 2 3 4 5
• Involves the team in planning the project	1 2 3 4 5
• Effectively communicates project status on an ongoing basis	1 2 3 4 5
• Brings a sense of excitement to the project	1 2 3 4 5
Encourages contrary opinions	
• Values a range of people's styles, talents, and opinions, even when they are markedly different from his/hers	1 2 3 4 5
• Actively listens to diverse points of view	1 2 3 4 5
• Is tactful and not offensive, even when disagreeing	1 2 3 4 5
Establishes a vision	
• Establishes and communicates clear and motivating objectives	1 2 3 4 5
• Conveys a strong sense of mission by enthusiastically expressing sense of direction and purpose	1 2 3 4 5
• Detects and focuses on important points in complex situations	1 2 3 4 5
Takes risks	
• Likes to take on new projects and programs	1 2 3 4 5
• Makes it safe for team members to take reasoned risks and honors those who do	1 2 3 4 5
• Willing to take risks on promising, unproven approaches	1 2 3 4 5
Creates positive environment	
• Relates well to individuals at all levels of the organization	1 2 3 4 5
• Provides encouragement and support to others for individual growth	1 2 3 4 5
• Shows empathy for others	1 2 3 4 5
• Appropriately applies a good sense of humor	1 2 3 4 5
• Uses personal management style to put others at ease	1 2 3 4 5

continued on next page

Figure 1-2   *Leadership Assessment (continued)*

Leadership characteristic	Rating				

Challenges existing beliefs

• Eliminates routines that don't serve a purpose	1	2	3	4	5
• Thinks and acts in fresh, creative ways	1	2	3	4	5
• Encourages team members to think creatively	1	2	3	4	5
• Questions assumptions and the status quo	1	2	3	4	5

Chooses his/her reactions

• Chooses to focus on the issue/situation rather than blame an external factor or get upset	1	2	3	4	5
• Focuses on the positive	1	2	3	4	5
• Remains calm, even in difficult situations	1	2	3	4	5

Recognizes performers

• Provides prompt, positive, and specific recognition for a job well done by personally thanking the person	1	2	3	4	5
• Gives constructive feedback in a way that benefits individuals when needed to resolve a performance issue	1	2	3	4	5
• Celebrates accomplishments of others	1	2	3	4	5

Is decisive, competent

• Makes realistic decisions	1	2	3	4	5
• Makes timely decisions	1	2	3	4	5
• Gets the team members the resources they need to succeed	1	2	3	4	5

Aligns individual and project goals

• Fully utilizes team members' skills, expertise, and energy	1	2	3	4	5
• Honors the opinions of others regardless of their status or position	1	2	3	4	5
• Delegates effectively to promote growth of team members	1	2	3	4	5

Establishes and achieves doable goals

• Divides projects into well-defined, doable tasks	1	2	3	4	5
• Sets clear priorities for self and others	1	2	3	4	5
• Gives the team members feedback as their tasks progress	1	2	3	4	5
• Establishes a clear plan and follows through	1	2	3	4	5
• Perseveres even in difficult circumstances	1	2	3	4	5

Figure 1-3   *Leadership Assessment—Rating Summary Form*

Leadership characteristic	Observers' Ratings										Avg	Self
	Obs 1	Obs 2	Obs 3	Obs 4	Obs 5	Obs 6	Obs 7	Obs 8	Obs 9	Obs 10		
Is honest, establishes trust • Follows through on all commitments; does what he/she says he/she is going to do												
• Sets a personal example of what he/she expects from others												
• Makes beliefs and values known, and lives them												
• Admits personal mistakes, learns from them, moves to correct the situation												
Gets people involved • Is open, approachable; keeps in touch with team members												
• Involves the team in planning the project												
• Effectively communicates project status on an ongoing basis												
• Brings a sense of excitement to the project												
Encourages contrary opinions • Values a range of people's styles, talents, and opinions, even when they are markedly different from his/hers												
• Actively listens to diverse points of view												
• Is tactful and not offensive, even when disagreeing												
Establishes a vision • Establishes and communicates clear and motivating objectives												
• Conveys a strong sense of mission by enthusiastically expressing sense of direction and purpose												
• Detects and focuses on important points in complex situations												
Takes risks • Likes to take on new projects and programs												
• Makes it safe for team members to take reasoned risk and honors those who do												
• Willing to take risks on promising, unproven approaches												

continued on next page

Figure 1-3 *Leadership Assessment—Rating Summary Form (continued)*

Leadership characteristic	Observers' Ratings										Avg	Self
	Obs 1	Obs 2	Obs 3	Obs 4	Obs 5	Obs 6	Obs 7	Obs 8	Obs 9	Obs 10		
Creates positive environment • Relates well to individuals at all levels of the organization												
• Provides encouragement and support to others for individual growth												
• Shows empathy for others												
• Appropriately applies a good sense of humor												
• Uses personal management style to put others at ease												
Challenges existing beliefs • Eliminates routines that don't serve a purpose												
• Thinks and acts in fresh, creative ways												
• Encourages team members to think creatively												
• Questions assumptions and the status quo												
Chooses his/her reactions • Chooses to focus on the issue/situation rather than blame an external factor or get upset												
• Focuses on the positive												
• Remains calm, even in difficult situations												
Recognizes performers • Provides prompt, positive, and specific recognition for a job well done by personally thanking the person												
• Gives constructive feedback in a way that benefits individuals when needed to resolve a performance issue												
• Celebrates accomplishments of others												
Is decisive, competent • Makes realistic decisions												
• Makes timely decisions												
• Gets team members the resources they need to succeed												

continued on next page

Figure 1-3 *Leadership Assessment—Rating Summary Form (continued)*

Leadership characteristic	Observers' Ratings										Avg	Self
	Obs 1	Obs 2	Obs 3	Obs 4	Obs 5	Obs 6	Obs 7	Obs 8	Obs 9	Obs 10		
Aligns individual and project goals • Fully utilizes team members' skills, expertise, and energy												
• Honors the opinions of others regardless of their status or position												
• Delegates effectively to promote growth of team members												
Establishes and achieves doable goals • Divides projects into well-defined, doable tasks												
• Sets clear priorities for self and others												
• Gives the team members feedback as their tasks progress												
• Establishes a clear plan and follows through												
• Perseveres even in difficult circumstances												

Figure 1-4   *Observer Feedback for Chris Jones*

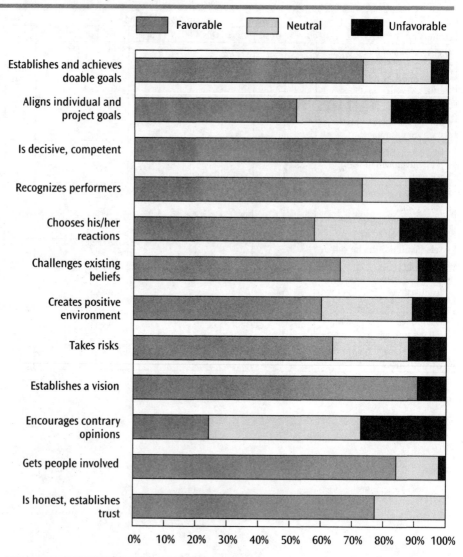

Develop a clear picture of your leadership characteristics as seen by your observers by creating a similar graph. This will help you complete Figure 1-5.

Figure 1-5   *Leadership Enhancement Worksheet*

Describe three of your leadership characteristics that both you and others perceived as strengths.

1. _____

2. _____

3. _____

Describe three of your leadership characteristics that are perceived more favorably by others than you perceive them yourself.

1. _____

2. _____

3. _____

Describe three of your leadership characteristics that are perceived less favorably by others than you perceive them yourself.

1. _____

2. _____

3. _____

List action steps to improve your leadership characteristics.

1. _____

2. _____

3. _____

4. _____

5. _____

6. _____

Figure 1-6   *Checklist of Ideas for Leadership Improvement Activities*

(Also refer to checklists at the conclusion of Chapter Two)

☐ Think about your leadership goals when you create your to-do lists. Schedule at least one activity each week that is specifically designed to help you meet these goals.

☐ Find a leadership mentor and explain your goals to the mentor. Meet regularly with the mentor to discuss your progress and solicit ideas for development of your leadership skills.

☐ Find someone you admire as a leader and ask that person to share his or her thoughts on leadership with you, including what it takes to be a good leader and why some succeed or fail as leaders.

☐ Look for opportunities to practice leadership outside your work environment such as in civic or volunteer organizations. Many chambers of commerce have active, local leadership programs. Look into becoming involved in such a program.

☐ Identify tasks that you must do yourself and tasks you can delegate. Review Chapter Five for information on effective delegation.

☐ Set specific self-improvement goals for yourself. Develop a plan to achieve these goals that identifies what it will take to achieve them, the resources needed, and where you will find the time to do so.

☐ Find opportunities to practice giving feedback, such as coaching a sports team or teaching a class.

☐ Start keeping a file of new ideas that appeal to you, even though they may not appear to be relevant to any of your immediate activities.

☐ Practice public speaking at every opportunity. Take a presentation course. Join Toastmasters.

☐ Attend presentations and lectures by others. Note what works or doesn't work for them in terms of connecting with the audience and getting the message across.

☐ For the next two weeks, commit to using the word "we" instead of "I." Leaders don't get the job done alone. Using "we" communicates a commitment to teamwork.

☐ Share information on the project status on a weekly basis. The team wants to know how they are doing.

☐ Keep track of how you are spending your time. What is the balance between the time you are spending on the *Content*, *Procedural*, and *Relationship* sides of the Triangle of Needs?

☐ Choose a famous leader whom you think is an effective role model, and read a biography or watch a movie about that person.

☐ Watch a video of Dr. Martin Luther King's "I have a dream" speech. This brief speech had a tremendous national impact and inspired many to advance his ideals. Identify the techniques he used to create such a powerful vision in such a short speech. Identify how you can use some or all of these techniques in communicating your vision. Even though you may not have the oratory skills of Dr. King, you can use the same techniques to make the appeal of your vision clear to others.

☐ Ask people how they would like to be recognized for their accomplishments, and follow through by providing the recognition.

☐ Take courses to become more creative, such as classes in acting, singing, improvisation, painting, or other areas of performance and art.

☐ Learn something entirely new and challenging, such as skiing or a foreign language.

☐ Do something dramatic to underscore your values. For example, if you value creativity and are stressing that your team come up with some creative solutions to an issue on your project, take them to a toy store and have them play with toys for an hour. Then, discuss what they learned about creativity that could be applied to the project issue.

☐ Read one or more of the following books and identify leadership practices that you want to apply. Pick the most important practice and create a plan to implement that concept. Once that practice becomes second nature, move on to another one on the list.

Principle-Centered Leadership, by S.R. Covey. Fireside Press, New York, 1992.

Leadership Is an Art, by M. Depree. Bantam Doubleday Dell Publishing Group, Inc., New York, 1989.

Peak Performers: The New Heroes of American Business, by C. Garfield. William Morrow and Co., New York, 1986.

The Leadership Challenge (3rd Ed.), by J.M. Kouzes and B.Z. Posner. Jossey-Bass, Inc., San Francisco, 2002.

Seeds of Greatness, by D. Waitley. Simon & Schuster, New York, 1983.

The Leader in You, by D. Carnegie and Associates. Simon & Schuster Pocket Books, Old Tappan, N.J., 1993.

The Seven Habits of Highly Effective People, by S. R. Covey. Simon & Schuster, New York, 1989.

Learning to Lead: A Workbook on Becoming a Leader, by W. Bennis and J. Goldsmith. Basic Books, Perseus Books Group, Cambridge, Mass., 2003.

TWO

How Can We All Pull Together?

Valuing Individual Differences to Improve Team Performance

Individual Differences and Team Performance

When assembling a team, a project leader considers each individual team member's personality, personal behavioral style preferences, technical strengths, and personal developmental goals. There are some interesting similarities between project leaders and dogsled racers, who also carefully consider individual differences in assembling their teams. They look for lead dogs with both emotional and physical stamina coupled with intelligence. Lead dogs must be willing to be chased by teammates (much like a project leader who may be "chased" both by team members and customers). Some dogs find this less stressful than others. Although personal style preferences do not preclude anyone from carrying out a project task, an individual will find some types of tasks more stressful than others, depending on his/her personal style preferences.

A project leader is sensitive to situations that may be stressful to individual team members. Dogsled racers also have found that dominant dogs do not always make the best leaders (much like a project manager who doesn't listen to the team or the customer and is unwilling to adjust his/her own behavior to build relationships with team members). On dog teams, two point dogs follow directly behind the lead dog (much like assistant project managers) to gain experience so that they someday may be lead dogs and to respond to voice commands if the lead dog isn't listening (just in case the project manager doesn't hear what the customer is saying, for example). The dogs directly in front of the sled are called "wheel dogs" and are chosen for strength, stamina, and quick reflexes. They are usually younger dogs that need more maturity before they are placed in positions that have more of an effect on the team's direction (much like the younger project team members who play an important role by carrying out specific project tasks but need some seasoning before leading a team). On human teams, individual differences are also among the factors leaders consider in structuring the team and in adjusting their own behavior in the day-to-day interactions with the team.

A wide range of project problems result from individual personality differences that create tensions and misunderstandings. An effective leader knows that team success is primarily tied to a demonstrated understanding *and valuing* of these differences and the development of constructive relationships between team members. Just like an effective dogsled team, every team member knows where he/she fits. Individuals' assignments on the team make the most of their individual strengths with an eye on developing them to be future leaders. This chapter discusses some of the psychological factors and individual preferences for behavioral styles underlying human differences, and it describes how a project leader's understanding and valuing of these preferences can improve team performance.

Use this chapter to determine your style preferences and develop an action plan for improving team relationships and your leadership effectiveness.

What Makes Us Tick: Type Theory

A person's personality is composed of many psychological facets that are interrelated in a complex and unique way. Although it is not possible for any one analytical tool to capture all of these many facets, there is a widely used tool that identifies different individual preferences for taking in information, organizing that information, reaching conclusions, and dealing with the outer world. These are all characteristics that are particularly relevant to project team and individual performance.

The noted psychiatrist C.J. Jung developed a theory that there are predictable differences in the way people prefer to take in information, organize the information, and reach conclusions. When people differ in what they perceive and in how they reach conclusions, they will differ in their interests, reactions, values, motivations, and skills. These personal style differences, when not understood and valued, have caused major problems for many a team.

Isabel Myers and Katherine Briggs began in the 1930s to observe and develop ways to measure the differences cited by Jung. They designed a psychological instrument that explains, in scientifically rigorous and reliable terms, individual preferences according to Jung's theory. Their work resulted in the Myers-Briggs Type Indicator (MBTI*) instrument, a questionnaire consisting of

---

*MBTI and Myers-Briggs Type Indicator are registered trademarks of Consulting Psychologists Press, Inc.

multiple-choice questions that can be completed in 15 to 20 minutes. More than two million people complete the MBTI each year, making it one of the most widely used psychological instruments (Myers, 1998; Myers, McCauley, Quenk, and Hammer, 1998; Myers and Kirby, 1994).

Although the information gained from the MBTI goes far beyond what can be covered here, our goal is to show how even the basic information we present can be used to improve individual project leadership skills and team performance.

Value of Understanding Individual Preferences

Although human personality is far too complex to be described solely by the style preferences that are identified by the MBTI, understanding these preferences can provide the following valuable insights into organizational and project team issues (Culp and Smith, 2001; Hirsch and Kise, 2000):

- Understand how your preferences affect your leadership effectiveness.
- Understand how individuals prefer to take in information and to prioritize that information to make judgments.
- Identify potential blind spots or areas of vulnerability.
- Determine how best to present information to and communicate with the customer and team members.
- Understand normal differences among people in areas such as communication, time management, the work environment, best supervision style, preferred learning style, and many others.
- Demonstrate the value of diverse styles.
- Enhance problem solving and decision making.
- Improve effectiveness in dealing with conflict.
- Reduce stress by understanding which situations will energize an individual and which will stress that individual.
- Achieve more effective meetings by structuring and conducting them to meet differing individual needs.
- Improve productivity by making assignments that allow each person to work according to his/her own personal style preferences and to understand how others prefer to work.
- More effectively meet deadlines by understanding how different types deal with time.
- Help people better understand themselves and enhance their relationships with others.

Leaders develop a deeper awareness of their own leadership strengths and potential developmental needs by understanding their own style preferences. Learning how to access and appropriately express different style preferences is crucial to building relationships, communicating effectively, and promoting synergy among team members. Leaders learn to express a range of style preferences and are open to learn new ways of behavior and expression to improve individual relationships.

Although there are many good uses, there is one thing for which knowledge of an individual's preferences should not be used: prejudging a person's ability to perform a task. There is nothing in a person's style preferences that need be a constraint to performing a task. A certain task may be more stressful for some, depending on their preferences, but a style preference does not limit ability.

Style Preference Scales

The MBTI identifies personal style preferences on the following four scales that have two opposite preferences defining the extremities or poles of each scale:

- Orientation—Whether an individual prefers to focus attention inward or outward to get energy (Introversion or Extraversion)
- The perception function—The way an individual prefers to take in information and perceive the world (Sensing or Intuition)
- The judgment function—The way an individual prefers to make decisions (Thinking or Feeling)
- The function an individual prefers to use when interacting with the external world (Judging or Perceiving)

Psychological type is the underlying personality pattern that results from the dynamic interaction of these preferences.

It is important to realize that the above preferences are only preferences. A useful analogy is right-handedness versus left-handedness. If you have a preference for right-handedness, it does not mean that you never use your left hand. Although there may be instances where you will use only your right hand (e.g., writing a letter longhand or throwing a baseball), there may be situations where both hands are equally important (e.g., typing a letter or hitting a baseball). You may strongly prefer to use your right hand or you may prefer it hardly at all if you border on being ambidextrous. The same is true for the MBTI type preferences listed above. According to Jung's theory, everyone has a natural preference for one of the two poles on each of the four preference scales. A person may use both poles at different times, but not both at once and not with equal confidence. There is one pole that a person prefers, and, when using it, the person generally feels more at ease, competent, and energetic.

It is critical to recognize that *there is no right or wrong* to these preferences. Each identifies normal and valuable human behaviors. Each has it strengths and each has its own potential blind spots. The best baseball teams have a mixture of right- and left-handed batters and pitchers. A study found that project teams with complementary preferences for taking in information and for making judgments outperform teams where all the team members have the same preferences (Blaylock, 1983).

There are characteristics exhibited for each of the poles of the different preference scales when people are doing things in their most natural way, outside of any roles that they may have to play in an assignment or a job. For simplicity, the preference for either pole of a scale determines how we discuss or refer to an individual who favors that preference; for example, those with a preference for Extraversion are Extraverts, those with a preference for Introversion are Introverts, etc. Keep in mind that everyone uses both poles at different times. This shorthand for preferences is for convenience and does not imply that anyone exclusively uses only one pole of a preference scale.

Style Preferences and Leadership

Leadership is about behavior. The way we think, the nature of our perceptions, and the way others perceive us are all important variables in understanding leadership. Leadership is a psychological process, so leaders need to understand the patterns in their current behavior and how these patterns affect their interactions with others; and they need to recognize and employ the range of behaviors available to them. Leaders are sensitive to the consequences of their own behavior on those around them and realize the importance of making conscious choices about how they behave. The behavioral style preferences described by the MBTI offer valuable insights into your behavioral style preferences, what you need to do to broaden your range of behavior, and when your other, less-preferred styles should be used.

Completing an MBTI questionnaire and discussing the results with a qualified MBTI practitioner will provide you with the most complete and most accurate description of your personal style preferences. However, if you have not had the chance to complete a questionnaire, you will get some ideas about your preferences as you read this chapter. To find a local practitioner, go to the Association for Psychological Type (APT) Web site (www.aptinternational. org/membership/).

Where We Direct Our Energy: The Extraversion (E)–Introversion (I) Preference

Have you ever been to a meeting where some team members dominated the discussion while others didn't speak up but later voiced contrary opinions or offered valuable input?

The Extraversion–Introversion preference scale describes the orientation and direction of energy. People may direct their energy and attention primarily to the external world (Extraverts) or to their inner world of ideas, values, and experience (Introverts). It is unfortunate that contemporary use of these terms may improperly imply differences in social ability. This MBTI scale does not describe social ability; rather, it describes where people get their mental energy: when they are participating outwardly in activities or discussions, or when they are reflecting inwardly on information, experience, or ideas.

People with a preference for Extraversion are energized by interacting with others. They love to have a sounding board and prefer to bounce ideas off others and talk things out. Extraverts may prefer to have a radio or television playing in the background while they are working. They are usually perceived as easily approachable and gregarious. Extraverts prefer to generate their ideas in groups rather than by themselves. They may become drained if they spend too much time in reflective thinking without being able to bounce their ideas off others. They like the opportunity to express their thoughts and may become frustrated if they aren't given a chance to voice them.

People with a preference for Introversion usually work best and are energized when they have quiet time to think things through. Introverts prefer to think things through before saying them and wish that others would do the same. They prefer to have a quiet workplace. They may be perceived as good listeners but also may be perceived as distant and hard to get to know. They usually find meetings or parties to be an energy drain and need some quiet time to recharge their energy. Introverts prefer to state their thoughts without being interrupted. They process their thoughts internally and dismiss many of them as of no interest to others. Because of their internal processing of ideas, Introverts may reach a conclusion without discussing the thought process that got them to that conclusion.

The general characteristics and stressful situations for each preference are summarized in the checklist on page 41.

As you read the characteristics of opposing poles of the Extraversion–Introversion scale, and each of the other preference scales, you may find some characteristics associated with both poles that seem to fit you. This is not unusual. If the characteristics of only one pole of the preference scale seem to fit you, you have a very clear preference for that pole. If half the characteristics of one pole and half of the other pole seem to fit, then your preference is less clear, although you probably have a slight preference. The preceding text and the text that follows the checklist will give you some additional clues about your preference. Leaders recognize that both poles of each scale describe valuable characteristics of leaders and that they can move along the preference scale toward either pole as circumstances require.

Check the boxes that most describe your characteristics and stressful situations to get an idea of whether Extraversion or Introversion is your dominant preference.

Characteristics of Extraverts	Stressful situations for Extraverts
☐ Prefer to communicate by talking	☐ Working alone
☐ Work out ideas by talking them through	☐ Having to communicate by e-mail
☐ Learn best through doing or discussing	☐ Working for lengthy periods with no breaks or interruptions
☐ Are sociable and expressive	
☐ Readily take initiative in work and relationships	☐ Having to focus in-depth on only one project task
☐ Think out loud	☐ Getting only written feedback on project performance
☐ Mix well at social events	

Characteristics of Introverts	Stressful situations for Introverts
☐ Prefer to communicate in writing	☐ Working face-to-face with others for prolonged periods
☐ Work out ideas by reflecting on them	
☐ Learn best by reflection, mental practice	☐ Interacting with others frequently, in person or on the phone
☐ Focus in-depth on their interests	
☐ May be seen as private and contained	☐ Having to act quickly without time for reflection
☐ Take the initiative when the situation or issue is very important to them	☐ Having too many concurrent tasks and demands
☐ Think, then talk	
☐ Are quiet and reserved at social events	☐ Getting frequent verbal feedback

A person's Extraversion–Introversion preference is one of the easiest to determine: merely by interaction with the person. The above characteristics usually are evident after a relatively brief time of interaction. Based on several thousand applications of the MBTI, the U.S. population is equally divided between Extraverts and Introverts.

Those preferring Extraversion are drawn toward people and things outside themselves and tend to actively pursue external interaction, drawing mental and emotional energy from these exchanges. Extraverted leaders tend to initiate contact and seek out others, to be action oriented, and to like processing their thoughts out loud. Those preferring Introversion tend to direct their energy and attention toward reflection and to draw energy from quiet, introspective time. Introverted leaders tend to prefer to receive information in writing and then to have time to process it internally before making a decision.

Both of the Extraversion and Introversion preferences are useful to leaders and also can cause blind spots. As is the case for all preferences scales, leaders increase their effectiveness by consciously directing energy to alter their behavior to compensate for their blind spots.

The Extraversion and Introversion preferences bring a leader the following strengths and weaknesses:

Project leader style	Strengths	Potential weaknesses
Extraverted	• Readily engage team members in discussion of ideas • Stay in touch with team members • Appear accessible to team members • Are more likely to show their enthusiasm and excitement	• May take quick action before giving enough internal reflection • External processing of ideas can be confusing to the team • May not give introverted team members enough time for internal processing • May appear inconsistent to those who don't understand their tendency to think out loud
Introverted	• Seen as good listeners and eager to receive information • Appear calm and focused • Think things through before acting • Deliver well-thought-out, consistent information to the team	• May continue to think when it is time to act • Internal processing may exclude others who want to participate in the process • Eventual announcement of a decision may seem to come out of the blue to the team because the decision-making process was internalized • May be seen as reserved or aloof, causing team members to feel excluded

Leaders value the needs of Extraverts on their teams by giving them an opportunity to talk things through and get issues out of their systems, giving them feedback about their performance. Engaging them through interaction provides a process for the Extraverts to develop their thoughts.

Leaders value the needs of Introverts by giving them time for thought, asking about their ideas, and allowing a slower pace for their reactions. Adjusting the pace of the conversation improves the quality of what is shared as well as the feeling of openness.

The Extraversion–Introversion differences can lead to tension between the team leader and the team and between individual team members. Extraverts may often invade the quiet time that Introverts need to think things through, but Extraverts may find their thought process is inhibited unless they can talk things through with someone. An Extravert project leader may be merely thinking out loud about several different ideas, but the Introvert team member may believe that the Extravert leader must certainly have thought it through before speaking. As a result, the Introvert may take each idea as something that is to be pursued during the project. This can lead to team members going off on tangents that may not advance the team toward the goals of the project. In this situation, Extravert leaders can help by saying that they are just thinking out loud and by being clear when they are stating a conclusion rather than a thought.

On one team that we worked with, there were sixteen Introverts (including the team leader) and two Extraverts. The Extraverts were seen by the others as disruptive to team meetings because they were perceived by the Introverts as shooting from the hip and not sticking to the points on the agenda. The entire team took the MBTI and gained an understanding of the Extraversion–Introversion differences. The Extraverts now preface any out-loud thinking by reminding the other team members that they need to throw out some ideas, and they ask assistance from the group in thinking through the ideas. Recognizing their differences has contributed to a more cohesive team.

Because an Introvert likes time to think things through, it is not productive to give an Introvert a lot of new information and ask for an immediate response. Often, when not given time to think things through, an Introvert's initial response will be negative to new ideas or suggested changes. Give an Introvert time to think over a suggested change or review a draft report before asking for his/her response. Introverts may be more comfortable giving their responses in writing.

Project teams sometimes engage in brainstorming sessions to seek creative solutions to project issues. Without some controls, the Extraverts will quickly dominate such a session and much potentially valuable input from the Introverts may be lost. To avoid this, the project leader should allow a few minutes at the start of the brainstorming session for each participant to write down his/her own ideas on the issue before the group. Then, take one idea from each person and write it on a flip chart before the entire group. Continue taking one idea per person until each person has exhausted his or her individual list. Do not allow any discussion of the ideas until all of the ideas have been listed. Then clarify and discuss each idea. This approach equalizes participation by providing the Introverts some time to think about their contributions and allowing the Extraverts the opportunity to discuss their ideas.

To become a leader, both Extraverts and Introverts must monitor their own preference as well as understand others' preferences. Introverts must recognize that sooner or later (preferably sooner) they must stop thinking and start talking. Introverts need to work on sharing their ideas more quickly and should not rule out any of their ideas or thoughts as too trivial too bring up. They should not hold others to the first words out of their mouths because an Extravert may just be thinking. Introverts are less likely to give praise to team members and must make a conscious effort to do so. Extraverts must recognize that sooner or later (preferably sooner) they must stop talking and start listening. Extraverts need to make a special effort to listen, avoid interrupting, and not assume that others' pauses in the discussion are an invitation to jump in. Introverts may be pausing to reflect on what they are saying before continuing.

Leaders draw upon their knowledge of style preferences to understand and respond to team members. When presenting information to the team, the leader expects the Extraverts to react immediately and to become clearer about their ideas as they talk them through. It is unlikely that the Introverts will express their opinions unless they had written materials in advance with time to reflect upon them. Extraverted leaders—in their desire for response and dialogue—need to be cautious not to fill the space with more words when the Introverts need time to reflect. Introverted leaders need to be prepared for the immediate questions and responses of Extraverts and not feel that they are being interrupted or treated rudely.

Figure 2-1 appears on page 61

Now turn to the Extraversion–Introversion Practice Checklist (Figure 2-1) and pick one action step that you will practice and use to help you work more effectively with team members who have the opposite preference. Write that action step in the blank below.

My preferred style is: The E–I action step I will take is:

☐ Extraversion (E)

☐ Introversion (I)

How We Gather Information:
The Sensing (S)–Intuition (N) Preference

Have you seen tension between some people who seem to focus on the big picture and others who seem to focus on the details, and they seem to talk past each other or stop in frustration because the other person just isn't "getting it"?

The Sensing–Intuition preference scale describes two ways of gathering information. Those who prefer to gather information through Sensing focus on data that are available to the senses, that is, the details and what is actual in the present time. They focus on practical realities. Those who prefer to gather information through Intuition focus on the connections and patterns of the data, looking for how the details fit into a bigger picture.

People with a preference for Sensing prefer to take in the details of information that is real and tangible. They tend to be very observant about the specific details of what is going on around them and focus on practical realities. A Sensor prefers specific answers to specific questions. They like to concentrate on the task at hand and find most satisfaction in tasks that yield a tangible result. They are more comfortable and energized by working with facts and figures than theories. They are more likely to be comfortable working on the detailed tasks of a project than they are on the conceptual planning of the project. They prefer clear project task descriptions rather than getting an overall plan with the details to follow. Sensors like to hear about things in a logical sequence rather than randomly. They would rather be doing something than thinking about it.

People with a preference for Intuition like to take in information by looking at the big picture. They focus on the relationships among the detailed facts and look for patterns and new possibilities. Intuitives tend to think about several things at once. They find the future possibilities to be exciting and intriguing, more so than current information. Intuitives like to figure out how things work just for the pleasure of it. They look for connections behind things rather than accepting them at face value. Intuitives tend to give general directions and may get irritated when a team member pushes them to be more specific. They are more likely to be energized by planning the project than doing the detailed tasks.

The general characteristics and stressful situations for each preference are summarized below. Check the boxes that most describe your characteristics and stressful situations to get an idea of whether you have a preference for Sensing or Intuition. The preceding text and the text that follows the checklist will give you some additional clues about your preference.

Characteristics of Sensors	Stressful situations for Sensors
☐ Focus on present realities and concrete facts	☐ Having to change the way things have been successfully done on past projects
☐ Focus on what is real and actual	☐ Having to give an overview without knowing the details
☐ Observe and remember specifics	
☐ Build carefully, step-by-step toward conclusions	☐ Talking about the big picture before understanding the detailed facts
☐ Understand theories through practical application	☐ Focusing on tomorrow's possibilities rather than today's tasks
☐ Prefer the certain outcomes that will result from past experience	☐ Dealing with people who are always coming up with new ideas
☐ Are often viewed as very practical, matter-of-fact	
☐ Are most energized when carrying out tasks involved in a project	

Characteristics of Intuitives	Stressful situations for Intuitives
☐ Oriented toward future possibilities	☐ Having to do things because it is the way they were done on past projects
☐ Are imaginative and creative	
☐ Focus on patterns and meanings in data	☐ Having to attend to details
☐ Remember specifics when they relate to a pattern	☐ Checking the accuracy of facts
	☐ Focusing on past experience
☐ Want to clarify a theory before putting it into practice	☐ Having to deal with today's tasks without seeing how they fit into the big picture
☐ Are excited about the possibilities associated with unproven theories	
☐ Are often viewed as ingenious	
☐ Are most energized in the planning phase of projects	

About 73% of the U.S. population has a preference for Sensing. One's preference can be affected by a situation. For example, even a person with a very strong preference for Intuition must become a strong Sensor by April 15 of each year to deal with the specific facts and figures of income tax returns.

The Sensing and Intuition preferences bring a leader the following strengths and potential weaknesses:

Project leader style	Strengths	Potential weaknesses
Sensing	• Give substance to visions • See practical impacts of alternatives • See that all needed details are considered when making decisions • Focus only on the parts of systems that need changing, not on change for change's sake • Preserve traditions and resources during times of rapid change	• Fail to see all the interactions as part of a bigger picture • Have difficulty in seeing how new ideas will work before experiencing them • Change too little when change is needed • Hold on to the past when it is time to let go • Do not recognize the value and validity of intuitive insights, either their own or others'
Intuition	• Have great energy to explore new ideas • Envision future directions • Take action on their vision for the project with confidence • Recognize potential interactive effects of project decisions • Can persuasively present a picture of the future	• Go ahead confidently without enough specific information • Make changes that may destroy what is now working • Fail to see all the steps, time, and resources required to achieve the project vision and goals • Institute more changes on the project before previous ones have been integrated • Discount past experiences that may offer valuable insights • Are impatient with others who don't get the vision for the project

Leaders value the Sensors on their teams by being sure not to discount factual information, accommodating requests for practical direct experience, showing appreciation for attending to specific information, and following through by doing exactly what was promised.

Leaders value the Intuitives on their teams by inviting abstract and theoretical discussion; by tolerating what may appear to be an irrelevant conversation, recognizing that it may actually be the shortest path to a project goal; and by embracing thoughts of what "could be" as well as "what is."

Both Sensing and Intuition preferences have valuable contributions to make to project leadership. Sensing will cause the leader to press others to keep their ideas simple and to the point. Sensing will test the Intuitives' broad concepts against reality. Sensing will push for common sense in even the most complex problem-solving situations. Intuition will stimulate the leader's imagination to see how the facts of the situation may be tied together to reach a creative solution and to throw out alternatives that should be examined to be sure that the best solution is reached. Use of both Sensing and Intuition makes it more likely that both the forest and the trees will be considered.

The style differences between Sensors and Intuitives can also lead to project tensions. In one case, we saw an architect and a mechanical engineer at odds. The architect was suggesting some conceptual changes to the ceiling of a building to improve the aesthetics of the reception area. The mechanical engineer was saying that it would never work because the revised design was intruding on the space where he had installed heating ducts. The Intuitive architect was seeing the big picture, and the Sensing mechanical engineer was focusing on the details. The situation had degenerated to the engineer ridiculing the architect's concept and crying "scope change." The architect was criticizing the engineer for lack of imagination, rigidity, and lack of customer orientation. By helping them understand the differences in their natural tendencies, the project leader helped them to better understand each other's perspective and work together to make the detailed changes that would be needed to support the new concept.

Sensors and Intuitives will deal differently with a request for a preliminary estimate of cost on a project. Intuitives are likely to be comfortable giving an estimate based on experience with other similar projects. Sensors are likely to respond by asking for more details so that they can build an item-by-item estimate. The Sensor may even view the request as unanswerable because it calls for information that the Sensor does not have. In these situations, it is helpful for the Sensor to keep in mind that a ballpark request by an Intuitive can be responded to without being exact. The Intuitive should recognize that the specifics that a Sensor is seeking are an effort to obtain what is perceived to be necessary information.

The closeout phase of a project involves dealing with a lot of details, such as deciding which materials can be discarded and seeing that all relevant materials are collected for storage and not scattered among individual files. Although all may have the same ability to do these tasks, it will likely be much less stressful to deal with them for a Sensor than for an Intuitive. The project leader will adjust assignments by considering the preferences of the individual team members.

The Sensing and Intuition preferences both play an important part in effective problem solving. First, leaders gather the facts by using Sensing to examine the details of the issue or problem. Then, leaders use Intuition to develop possible causes and solutions. This approach asks a person to use both the preferred and the nonpreferred styles. Because this is difficult for some, there is value in having individuals on the team that offer a mix of Sensing and Intuitive preferences.

Figure 2-2 appears on page 62

Now turn to the Sensing–Intuition Practice Checklist (Figure 2-2) and pick one action step that you will practice and use to help you work more effectively with team members who have the opposite preference. Write that action step in the blank below.

My preferred style is: The S—N action step I will take is:

☐ Sensing (S)

☐ Intuition (N)

How We Make Decisions:
The Thinking (T)–Feeling (F) Preference

Are projects being delayed because decision making is stalled:
some are concerned about the impacts the decision will have on people,
while others are emphasizing the tangible pros and cons,
and neither accepts the other's point of view?

The Thinking–Feeling preference scale describes two different ways of making decisions. In one way, cause-and-effect logic of a situation and impersonal decision-making criteria are considered (Thinking). In the other way, personal values and interpersonal consequences are the key decision-making criteria (Feeling). There is no intention to imply that Thinkers don't have feelings or that Feelers don't think. It is unfortunate that the term Thinking may be associated by many with intellect and the term Feeling associated with emotion. Both Thinkers and Feelers can be equally intellectual and emotional. The terms are used here to describe the preferred decision-making criteria and have nothing to do with intellect or emotion. Both Thinking and Feeling are rational processes used to make decisions. People with a Thinking preference look at the logical consequences of a decision. They objectively examine the pros and cons. They are energized by examining an issue to find what needs to be done so that they can resolve the issue. They like to find a standard or principle that applies to all similar situations. Thinkers tend to settle disputes based on what they believe is fair and truthful rather than on what will make people happy. They often appear to be calm and objective in situations where others appear upset. Thinkers don't mind making difficult decisions and think it is more important to be right than liked. They are impressed with logical and scientific arguments and use impersonal decision-making criteria.

Those with a Feeling preference consider what is important to them and to others who are involved. They mentally place themselves into the situation so that they can identify with others and make a decision based on their values about honoring people. They are energized by supporting others. Feelers consider that a good decision is one that takes into account impacts on others. Feelers give a lot of weight to how a decision will affect others. They prefer harmony over clarity and do not like conflict. They enjoy providing needed services to people. They will extend themselves to meet others' needs, even at the expense of their own comfort. Feelers prefer to use personal, value-related decision-making criteria. The use of these different decision-making criteria may lead to the same conclusion, but the paths to the outcome and their behavior are going to differ.

The general characteristics and stressful situations for each preference are summarized below. Check the boxes that most describe your characteristics and stressful situations to get an idea of whether you have a preference for Thinking or Feeling. The preceding text and the text that follows the checklist will give you some additional clues about your preference.

Characteristics of Thinkers	Stressful situations for Thinkers
☐ Are analytical	☐ Noticing and appreciating positive performance by others
☐ Use cause-and-effect reasoning	
☐ Solve problems with logic	☐ Focusing on processes and people
☐ Prefer impersonal, objective decision-making criteria	☐ Using empathy and personal values to make decisions
☐ Strive for an objective standard of truth	☐ Having others perceive their questions as divisive
☐ Can be perceived as strong-willed	
☐ Prefer to be viewed as fair-minded	☐ Taking time to consider individual reactions to decisions

Characteristics of Feelers	Stressful situations for Feelers
☐ Are empathetic	☐ Setting criteria and standards
☐ Use personal, value-related decision-making criteria	☐ Focusing on project tasks only to the exclusion of how they affect others
☐ Assess impacts of decisions on people	☐ Being expected to use logic alone to make decisions
☐ Strive for harmony	
☐ Are compassionate	☐ Critiquing others' work and looking for flaws
☐ Can be perceived as tender-hearted	
☐ Prefer to be viewed as caring	☐ Confronting performance issues

The Thinking and Feeling preferences are the only two preferences where there is any gender difference. Slightly less than two-thirds of males have a Thinking preference and slightly more than two-thirds of females have a Feeling preference. Overall, about 60% of the U.S. population has a Feeling preference.

The Thinking and Feeling preferences bring a leader the following strengths and potential weaknesses:

Project leader style	Strengths	Potential weaknesses
Thinking	• Analyze problems logically • Make decisions objectively • Make hard decisions in a timely way • Develop clear rationale for decisions • Are willing to move on and leave the past behind	• Difficulty in recognizing that logic does not persuade everyone • Difficulty in dealing with others' needs for time to process • Discomfort in dealing with others' feelings • Impatience with negativity • May not adequately consider reactions of and effects on individuals when making decisions
Feeling	• Include others in gathering information and making decisions • Appreciate the team members' contributions • Recognize the need for and provide individual processing time and support • Remember past contributions • Consider the impact of decisions on people	• Put excessive energy into inclusion, consensus building, and harmony • Fail to confront difficult people and difficult decisions • Focus on team harmony to the detriment of tasks that will allow the team to advance toward the project goals • Fail to see problems in individuals that they feel connected to, even though these individuals may be causing difficulties for the team • Romanticize the past

Leaders value Thinking members of their teams by giving them independence to pursue answers to questions, allowing the critique of thought and experience, and recognizing what seems to be critical: that questioning and analyzing behavior is an indication of the Thinker's interest. Leaders address the needs of Thinkers by presenting well-thought-out plans for the project, by demonstrating they are competent to lead the project, and by recognizing the Thinker's good ideas and work.

Leaders value Feeling members of their teams by focusing on values and human relationships, giving equal value in decision making to the people issues, and providing time for the Feeler to articulate how a project situation or decision may affect the quality of the team members' lives. Leaders take into account the needs of Feelers by openly discussing people issues related to the project and to the team. They demonstrate that they have thought about these issues and want to involve team members in addressing them. Leaders frequently and publicly express recognition and appreciation for team members' contributions.

The Edsel automobile, introduced by Ford Motor Company in 1958, is a classic example of what can happen when a project team is predominantly composed of Thinkers. The project engineers designed a car that was very advanced technically, with innovative features such as a push-button transmission and innovative styling. The Edsel project team was made up of Ford's best engineers, and they were confident that they had the best-engineered car on the market. Unfortunately, no one on the team considered the subjective (Feeling) impact of the car. How would people feel about a car that was so different? The fact that the car did not sell well showed that people's feelings about the Edsel were not good. Every project team needs a leader and enough members capable of using their Feeling preference to consider how the end user of their work will feel about the results.

Figure 2-3 appears on page 63 Now turn to the Thinking–Feeling Practice Checklist (Figure 2-3) and pick one action step that you will practice and use to help you work more effectively with team members who have the opposite preference. Write that action step in the blank below.

My preferred style is: The T−F action step I will take is:
☐ Thinking (T)
☐ Feeling (F)

How We Interact with the Outer World:
The Judging (J)–Perceiving (P) Preference

Are some people on the team always late for meetings,
causing others who always are on time to become irritated?
Do some people strive to a conclusion or closure regardless of changes,
while others resist ever reaching a conclusion because
they want to keep their options open?

The Judging–Perceiving preference scale describes whether a person prefers to use his/her decision-making, judging function (Thinking or Feeling) or information-gathering, perceiving function (Sensing or Intuition) when interacting with the outer world.

People who prefer to use their decision-making Thinking or Feeling function have a preference for Judging. People who prefer Judging like to live in a planned and orderly way. They want to make decisions, reach closure, and move on. Judgers tend to be organized and structured. They like to have things settled. Schedules are very important to them. Judgers do not like clutter and usually have a clean, orderly desk and office. They have a place for everything and are not happy until everything is in its place. They don't like surprises. They have a schedule and plan for their project work and may get agitated or frustrated if things do not go as planned. They tend to make thorough to-do lists. They start their project work early to avoid last-minute crunches that they find very stressful. It is difficult for a Judger to relax until the work is done.

People who prefer to use their information-gathering Sensing or Intuitive function when interacting with the outer world have a preference for Perceiving. People who prefer Perceiving like to live in a flexible, spontaneous way. Detailed plans and schedules feel confining to Perceivers. They prefer to stay open to ongoing information and last-minute options. They enjoy the process more than closure. Perceivers like to explore new ways of doing things. They often do not have a detailed plan for their project work but prefer to see what the project demands. Neatness is not as important as spontaneity and creativity. It is difficult for Perceivers to start tasks any sooner than they perceive is absolutely necessary because they are energized by the pressure of meeting a deadline. They have no problem relaxing first and doing the work later. Perceivers like to keep their options open and often avoid being pinned down. Time commitments are approximate, not absolute, and are easily subject to change if a Perceiver's priorities switch.

The general characteristics and stressful situations for each preference are summarized below. Check the boxes that most describe your characteristics and stressful situations to get an idea of whether you have a preference for Judging or Perceiving. The preceding text and the text that follows the checklist will give you some additional clues about your preference.

Characteristics of Judgers	Stressful situations for Judgers
☐ Scheduled	☐ Waiting for project structure to emerge from a general planning process
☐ Organized	
☐ Systematic	☐ Uncertainty about time frames and deadlines
☐ Methodical	
☐ Make short- and long-term plans	☐ Rushing at the last minute to meet a deadline
☐ Like to have things decided	
☐ Try to avoid last-minute stress	☐ Dealing with surprises

Characteristics of Perceivers	Stressful situations for Perceivers
☐ Spontaneous	☐ Having to plan ahead
☐ Flexible	☐ Working within time frames and deadlines
☐ Casual	☐ Having to finish and move on
☐ Open-ended	☐ Having others distrusting that their last-minute bursts of energy will meet the project deadline
☐ Adaptable	
☐ Like things loose, open to change	
☐ Energized by last-minute pressure	

About 54% of the U.S. population has a preference for Judging.

The Judging and Perceiving preferences bring a leader the following strengths and potential weaknesses:

Project leader style	Strengths	Potential weaknesses
Judging	• Set clear schedules and goals • Develop well-organized plans for project work • See that teams follow through on project plans and meet schedules • Strive for closure on even the most difficult of decisions	• Uncomfortable dealing with ambiguity and uncertainty • Difficulty in staying motivated in midst of confusion • Difficulty in adjusting project plans when circumstances change • Difficulty in trusting that Perceivers will get tasks done at the last minute • Desire for closure may lead to premature decisions
Perceiving	• Remain open to options and eager to explore them with team • Consider many options before making a decision • Adapt project plans as changes occur • Willing to revisit and revise previous decisions • Open to new information at any time during the project	• Difficulty in developing project plans and schedules • Desire to collect more information or consider more options when it is time to make a decision • Tendency to stretch budgets and schedules in order to collect more information or consider more options • Last-minute decisions may frustrate team members with Judging preference

The authors worked with one company where the staff was very frustrated by the lack of decision making by their management. The MBTI results showed that every member of the board had a Perceiving preference, an unusual situation as past studies have shown that 60% to 70% of managers have a Judging preference. By recognizing their tendency to get bogged down in their enjoyment of the process of discussing issues, the board restructured their meeting agendas to establish a time where closure of the process on each agenda item was to be achieved and a decision made. They also added a facilitator to their meetings to help keep them on track toward making decisions.

A situation that is ripe for problems is where a project manager has a Perceiving preference and is working with a customer that has a strong Judging preference. We witnessed the disastrous results once. The Perceiving project manager arrived at the first project meeting 20 minutes late, which a Perceiver may look at as "on time." The Judging customer viewed it as 20 minutes late and an insult. The customer bluntly told the project manager not to do it again. The project manager did do it again, and again, and the customer fired the company. In another similar situation, with a better outcome, the Perceiving project leader listened to the customer's complaint about tardiness and moved his departure time for meetings at the customer's office forward 30 minutes so that he was always a few minutes early rather than a few minutes late.

A leader who can use Judging and Perceiving preferences brings strength to a project team. Because of the desire for closure, Judging sometimes results in a desire by the leader to push to meet or beat deadlines even at times when, due to changed circumstances, the project deadline may not be the most important consideration. Using the Perceiving preference can offset this tendency by causing the leader to keep the project priorities open. The Judging preference assists by reminding the leader that there are schedules to meet and that project issues need closure. A leader uses enough Judging to be sure the team stays on course and enough Perceiving to be sure the team is not making good time going in the wrong direction.

Figure 2-4 appears on page 64

Now turn to the Judging–Perceiving Practice Checklist (Figure 2-4) and pick one action step that you will practice and use to help you work more effectively with team members who have the opposite preference. Write that action step in the blank below.

My preferred style is: The J–P action step I will take is:
☐ Judging (J)
☐ Perceiving (P)

How Style Preferences Affect Leadership Effectiveness

Figure 2-5 appears on page 65

Chapter One presents a Leadership Assessment tool that solicits the perceptions of others about an individual's leadership effectiveness. The authors have had the opportunity to administer both the MBTI and this Leadership Assessment tool during the course of conducting leadership workshops. The 360-degree feedback provided by the Leadership Assessment tool has provided an opportunity to match trends in how people perceive an individual's leadership effectiveness with the individual's MBTI behavioral preferences. We have developed mathematical correlations between Leadership Assessment ratings by others and the clarity of MBTI preferences to identify these trends. The results, and our thoughts on the underlying reasons for the correlations, are shown in Figure 2-5. These trends provide some insight into which behaviors are more likely to reinforce leadership and which may present some challenges.

No one is precluded from being an effective leader by his or her behavioral style preferences. Indeed, every preference appears in Figure 2-5 as both an asset or a challenge depending on the aspect of leadership considered, emphasizing the need to have the flexibility to use both preferred and less-preferred styles. The identification of a challenge for one of your preferences in Figure 2-5 does not mean that it will necessarily result in a low observer rating for the related Leadership Assessment category because you may have already consciously or subconsciously made adjustments in your behavior to address the challenge based on your experience in working with others.

Understanding these trends can assist you in understanding the underlying reasons for others' perceptions and in creating an action plan to improve your leadership effectiveness. For example, an individual in one of our leadership workshops received an especially low rating for "Establish and achieve doable goals" but a very high rating for "Establish a vision." Her MBTI results showed strong preferences for Extraversion, Intuition, and Feeling and a moderate preference for Perceiving (see "Establish a vision" and "Establish and achieve doable goals" in Figure 2-5 for a description of how these preferences may have affected her Leadership Assessment ratings). When we asked her if she had any comments about the observer ratings, she said the rankings made sense because she really enjoyed talking about and establishing big-picture objectives but did not enjoy working out a work plan or doing the detailed work. She recognized that she often delayed doing a project work plan. She went on to say that she had noticed that although she usually got her teams and customers excited about the potential end results at the *start* of her projects, some of her team members later seemed unclear about how they were to proceed with the project work, repeatedly coming to her to seek direction and wanting to talk about details that were annoying to her. She admitted this was no doubt frustrating for her team members as well. She observed that once the project was sold to the customer and the major objectives set, her interest level dropped.

After a discussion of her Myers-Briggs preferences, she quickly saw how her preferences for Intuition and Perceiving contributed to her behavior and led to the observer ratings. She commented, "This gives me a lot to think about; now I see what I have to work on." Recognizing how her own style preferences were contributing to the problem, she developed an action plan that focused on improving her flexibility to use Sensing and Judging. Realizing that all of her personal changes would not be made overnight, one element of her action plan was to assign the lead role in the development of a detailed work plan for a major project that was about to begin to a detail-oriented senior team member and then to work closely with that team member and the balance of the team in developing the plan in a timely way. Her goal was to use this as a personal learning experience and to demonstrate to her team that she was placing increased emphasis on developing a project plan that clearly defined not only the goals but also the means to achieve them.

The effects of telling her observers about her personal action plan and making visible progress went far beyond the direct and significant benefits of more effectively setting and achieving goals for the next project. She also will be viewed as more trustworthy, an essential characteristic of a leader, because her changes in behavior demonstrate that she listened to her coworkers and acted on their feedback in a visible way.

Keys to Effective Use of Psychological Type

The Myers-Briggs Type Instrument provides valuable insights into an individual's psychological type and personal preferences for taking in information, making decisions, and interacting with others. An understanding of their own and others' preferences allows leaders to broaden their own range of behaviors and to know when they should use their less-preferred styles in interactions with team members who have differing preferences. Understanding and valuing individual differences is the key to building strong relationships (Brock, 1994).

Leaders should keep the following aspects of psychological type in mind:

- The purpose of type theory is not to label people and put them into boxes to limit their possibilities. The purpose is to explain so that you understand yourself and others to improve relationships, not to confine.

- Type is for understanding, not for excuses. It should never be used to prejudge your own or another's ability to do anything. For example, saying you have a preference for Intuition and therefore can't be bothered by details is an excuse. It would be more useful to say that it is easier for you to grasp the details after you have the overall concept developed.

- Be aware of your biases toward preferences different from your own. For example, someone with a strong Judging preference may have a bias that those with a Perceiving preference are going to procrastinate. Remember that the Judging–Perceiving preferences are about whether an individual prefers to structure and schedule the decision-making process or leave it open for more information. Both approaches can bring value to a project.

- There are no right or wrong types. Each type brings its own strength to the project team and to other working relationships.

- Everyone uses all preferences to some degree. It can be helpful to view a situation through the eyes of your opposite preference or type. This will increase your familiarity with the way others may be thinking and will also increase your ability to call on your opposite preferences when a situation demands.

Personality is much more complex than these type preferences. Although very useful, type theory does not explain everything involved in working or personal relationships.

Trends in Preferences

In many cases, a majority of project team members may have the same preferences. This may be because they come from similar educational backgrounds or professions that attract individuals with certain preferences. As shown in the table below, there are clear trends for preferences in different professions (Myers, McCauley, Quenk, and Hammer, 1998; Culp and Smith, 2001).

Comparison of Type Preferences in Various Professions

	E	I	S	N	T	F	J	P
U.S. population	50%	50%	73%	27%	40%	60%	54%	46%
Engineering project team members	38%	62%	54%	46%	75%	25%	67%	33%
Mid- and upper-level managers	52%	48%	50%	50%	80%	20%	69%	31%
Long-term care nursing assistants	50%	50%	88%	12%	41%	59%	81%	19%
Male clergy	42%	58%	57%	43%	31%	69%	68%	32%
Pharmacists	38%	62%	74%	26%	63%	37%	67%	33%
Dental hygienists	57%	43%	75%	25%	43%	57%	73%	27%
Librarians	37%	63%	41%	59%	60%	40%	60%	40%
Life insurance agents	74%	26%	83%	17%	63%	37%	71%	29%
Basketball officials	66%	34%	97%	3%	NA	NA	81%	19%
Management consultants	58%	42%	33%	67%	62%	38%	59%	41%
Human resource personnel	59%	41%	38%	62%	61%	39%	61%	39%

E = Extraverted; I = Introverted; S = Sensing; N = Intuition; T = Thinking; F = Feeling; J = Judging; P = Perceiving; NA = not available.

For example, there is a very strong trend for Thinking and Judging preferences for engineers. Engineers who have migrated upward through their organizations to positions of project or company management will have to be especially aware of the effort it will take to use their Feeling and Perceiving preferences to move beyond being managers to become effective leaders.

MBTI Problem-Solving Model

Use of the Sensing–Intuitive and Thinking–Feeling preferences are an essential part of an effective problem-solving model illustrated below (Kroeger and Thusesen, 1992).

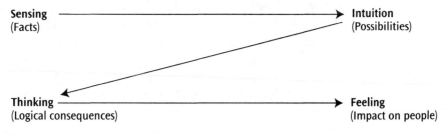

Sensing
(Facts)

Intuition
(Possibilities)

Thinking
(Logical consequences)

Feeling
(Impact on people)

Problem-solving model.

The Sensing preference is important in collecting all of the needed information and facts. Intuition is then useful in discerning patterns in the information and determining how the information fits into the big picture solution. The Thinking preference is then of value in considering the logical consequences of various alternative actions. The Feeling preference is then useful in weighing the impacts of these alternative consequences to consider how people will be affected. The process leads to a sound, well-balanced problem solution. Ideally, the leader will have the flexibility to use all of these preferences in solving problems. A team made up of a mixture of individuals with different preferences may also avoid a potential blind spot in the problem-solving process.

Balancing Preferences on Teams

Let's consider a project application of this problem-solving model. On a facility siting study, for example, the sites would typically be compared based on cost, location, access, etc. (facts, figures—Sensing). In addition, there would be discussions on the long-term feasibility, growth, and opportunities (possibilities, big picture—Intuition). At that point, there may be a clear winner when you weigh all of the pros and cons (logical consequences—Thinking), until you start to look at how the community feels about the facility and the potential influence/impacts (consequences—Feeling). Then the final decision making is usually a weighing of pros and cons and a balancing of the facts and figures with the values and social acceptance. You probably recognize the process, but it is vital to understand why it works; otherwise, there can be a lot of frustration when a location is picked that isn't "the cheapest," "most logical," or doesn't "consider the community's desires."

A leader who is aware of the team tendencies can include specific steps and assignments in the project approach to compensate for the potential vulnerabilities of the team. Let's consider an example of a hypothetical project team where the most common individual preferences are for Introversion, Sensing, Thinking, and Judging. This team may encounter problems such as the following:

- *Problem:* Relationships on the team and with the customer suffer because there are not enough face-to-face interactions (not using Extraversion).

  Potential solution: The project leader schedules a portion of his or her time each week to visit with each team member and take any appropriate actions based on what is learned. This will take a significant, conscious effort if the project leader also has an Introversion preference. Some options would be to schedule Friday lunches for the entire team or with selected team members, or to periodically schedule a lunch meeting with the customer and the team.

- *Problem:* The team members are focused on the details of the technical work and so focused on today's tasks and deadlines that they are blindsided by the unexpected. They view that it is more effective to be doing something useful today than to be wondering about tomorrow (using their Sensing preference).

Potential solution: Early in the project, the project leader can use the previously described brainstorming technique in a team meeting with the purpose of identifying problems that could arise on the project and how they could be addressed.

- *Problem:* The team members are so focused on the logical steps of the work (Thinking) that they overlook how the customer is going to feel about the end result. They may be building another Edsel!

Potential solution: The project leader can designate one or more team members to periodically meet with the customer. This meeting would be held solely for the purpose of asking how the customer is feeling about how the project is proceeding, not to discuss technical aspects of the work.

- *Problem:* The team members are so focused on reaching closure (Judging) that they don't take time to periodically step back and see if the situation is still the same or some new alternatives should be considered.

Potential solution: At selected milestones on the project, the project leader can have team members or individuals not otherwise involved in the project conduct a review of how work to date is aligning with the project goals and to discuss whether changed circumstances dictate a change in those goals.

Of course, each team will have its own unique set of type-related tendencies and solutions to consider based on the team members' characteristics and the project nature. For additional reading on applying the MBTI in the workplace, we suggest the book *Type Talk at Work* (Kroeger and Thuesen, 1992).

Exercises

1. You can enhance your leadership capabilities by improving your ability to access and use both ends of each of the style preference scales. By now, even if you haven't completed an MBTI questionnaire, you probably have a good idea of your preferred styles. Refer back to the items you checked in the characteristics and stressful situation checklists, and re-read the action steps you wrote at the conclusion of the discussion of each preference scale.

Figure 2-6 appears on page 69 Using Figure 2-6, mark the location along each scale to indicate the strength of your preference. For example, if you have a very strong preference for Thinking, you would place a mark very near the Thinking end of the Thinking–Feeling scale. If you feel that you do not have a strong preference, you would place a mark near the middle of the scale. If you have completed an MBTI question-naire, the results will identify the strength of your preference on each scale. Note the strength of your preference for each preference scale.

Use the "Action List" portion of Figure 2-6 to summarize the action steps from the practice checklists (Figures 2-1 through 2-4) that you noted as you read Chapter Two. Looking at the relative strengths of your preferences from the profile in the upper part of Figure 2-6, prioritize the scales and associated action steps. The highest priority should be assigned to the preference "stretch" that you believe would be of most benefit to you in becoming a more effective project leader. It will support you to discuss your action plan with peers and ask them to periodically check in with you to discuss your progress and to offer feedback.

As the highest priority action becomes more comfortable to you, move on to the next highest priority action step. Recognize that, at first, activities that emphasize your less-preferred styles may take a lot of energy to initiate and may be somewhat stressful. However, by consciously focusing time and energy on your less-preferred styles, it will become easier to access that style at times when it is important that you do so.

2. Review the results of the Leadership Assessment you conducted at the conclusion of Chapter One (Figure 1-2). Identify any connections between the discoveries you made there about your leadership strengths or areas for improvement and your style preferences. Would the action plan you created at the conclusion of Chapter One (Figure 1-5) benefit by including some of the practices shown in Figures 2-1 through 2-4? For example, if you have a preference for Perceiving and you found that others rated you low in regard to making timely decisions, selecting action steps to practice Judging may help you improve the timeliness of your decisions.

Figure 2-1   *Extraversion–Introversion Practice Checklist*

Extraverts can improve their flexibility to use Introversion by taking the following actions:

☐ Consciously pause when the feeling to take action is especially strong or when you are enthusiastic about a particular plan of action. Take time to reflect upon your ideas.

☐ Practice being silent for periods of time during meetings.

☐ Make a conscious effort not to interrupt others; let them finish speaking before you jump into the conversation.

☐ Stop, look, and listen.

☐ Recognize that sooner or later (preferably sooner) you must stop talking and start listening.

☐ Don't assume you can talk your way through and out of most conflicts.

☐ Before making a presentation, carefully think through the issues you want to cover and take time to rehearse.

☐ Pause and think before answering questions instead of saying the first thing that comes to mind.

☐ Postpone reacting to bad news.

☐ Practice contemplative behavior such as meditating, setting aside one hour of quiet time each day, keeping a journal, or talking long walks.

☐ Consciously focus on letting others speak first and being comfortable with periods of silence during a conversation.

☐ Actively listen to others to find out all you can about their skills and interests.

☐ Before jumping into a task, slow down and rethink what you are about to do.

☐ Ask team members and coworkers for their ideas—and listen.

☐ Give others time to think about your ideas before pushing them for feedback.

Introverts can improve their flexibility to use Extraversion by taking the following actions:

☐ Recognize that sooner or later (preferably sooner) you must stop thinking and start talking.

☐ Express yourself even if it means repeating yourself until you are sure you are heard.

☐ Engage in frequent one-on-one meetings to listen to team members and offer guidance and support.

☐ Meet a new business contact for lunch at least once per week.

☐ Solicit others' input even if you think you can do the task at hand alone.

☐ Sign up for a course or seminar that involves new groups of people.

☐ Take periodic walks around your workplace to talk to team members, and keep your office door open.

☐ Voice your thoughts in meetings, even if you wish you had more time to think them through.

☐ Make the effort to chat with others.

☐ Compliment others on their work.

☐ Be vocal and visible in your excitement.

☐ Get actively involved in a professional or trade association, such as Rotary clubs and Toastmasters.

☐ Attend drama classes offered by adult education programs.

Figure 2-2   *Sensing–Intuition Practice Checklist*

Sensors can improve their flexibility to use Intuition by taking the following actions:

☐ Take a class in strategic planning or long-range planning.

☐ Don't overwhelm others with facts.

☐ Relate facts to a bigger picture.

☐ Look beyond the facts to see what other issues need attention.

☐ Write a brief executive summary of your next report that describes the trends represented by the information in your report.

☐ Take courses that focus on plots or themes, such as courses in literature, art, or music.

☐ Look at abstract paintings or read books and consider what message the painting or book delivers rather than what details make up the painting or the book.

☐ Watch films that are abstract and symbolic, such as those directed by Berman or Fellini.

☐ Use quiet time to think beyond the present and focus on the future. Include regular time in your day or week to think about the medium- to long-range future rather than the issues of the day.

☐ Think about your personal "ideal scene" for the future. What do you want to be doing five years from now? How will you get there?

☐ Practice brainstorming with your team members.

☐ Learn new ideas by reading what successful organizations and teams do.

☐ Occasionally try new approaches just to gain the experience.

☐ Start keeping a file of new ideas that appeal to you, even though they may not appear to be relevant to any of your immediate activities.

☐ Write down any connections in your experiences by keeping a personal journal to record your observations, and consider the significance of connections between seemingly unrelated events or information.

Intuitives can improve their flexibility to use Sensing by taking the following actions:

☐ Stay focused on the present and what is happening right now.

☐ Present information in a step-by-step manner.

☐ Cook from recipes.

☐ Do things that require you to use your hands, such as plant a garden, paint, sculpt, or do some home repairs.

☐ Take stock of what your five senses are telling you. What do you see, hear, smell, taste, or feel by touch? Focus on enjoying things as they are.

☐ Demonstrate respect for organizational values and traditions.

☐ Focus on the reality of what is happening around you in its most concrete form.

☐ Practice expressing your thoughts in the most concise way using simple language. Provide specific examples on important topics.

☐ Practice expressing direct, specific facts.

☐ Learn the technical jargon used by your team.

☐ Resist the temptation to make changes for change's sake.

☐ Appoint or hire a detailed-oriented assistant, and learn from the way he or she assists you.

Figure 2-3 *Thinking–Feeling Practice Checklist*

Thinkers can improve their flexibility to use Feeling by taking the following actions:

☐ Focus on offering compliments rather than criticism, and comment on other's positive traits rather than point out what you view as negative traits.

☐ Remember that consideration of the impacts on people is logical, even if the people aren't always logical.

☐ Acknowledge your own emotions and feelings and express them, especially to others who have had similar experiences.

☐ Write about your values. Ask a close friend about his/her values.

☐ Spontaneously acknowledge the efforts of others with an appreciative comment or note.

☐ Work to gain the cooperation of others, rather than trying to force compliance. Ask yourself what is in it for them if they join up with you.

☐ Make a conscious effort to stretch the range of your normal conversational topics to share personal information about yourself.

☐ Look for points of agreement with others before discussing issues.

☐ Celebrate small successes with your team.

☐ Voice your confidence in others' abilities.

☐ Recognize people on significant dates, such as employment anniversaries or birthdays.

☐ Avoid stating your position in absolute terms.

Feelers can improve their flexibility to use Thinking by taking the following actions:

☐ Give direct, simple feedback to others. When you receive feedback, look for information that will be useful to you, and let any emotional energy in the feedback pass by. Pay attention to reacting less personally to criticism. Remember that feedback is often offered in the spirit of helping you improve.

☐ When faced with a business decision, list options and criteria to evaluate the options. Assign points for each option using the evaluation criteria, and select the highest scoring option.

☐ Practice making your points logically by considering causes and effects, perhaps using "if . . . then" language.

☐ Take a course in critical thinking.

☐ When leaving voice-mail messages or sending e-mail, see how concise you can make your message.

☐ Ask yourself cause and effect questions, such as, "If I say yes to this request, what does it do to my other plans?"

☐ Remember that some people are not seeking to turn their business relationships into friendships. Focus on the business at hand in these circumstances.

☐ Site specifics when you give others positive feedback.

☐ Recognize and promptly deal with performance issues.

Figure 2-4   *Judging–Perceiving Practice Checklist*

Judgers can improve their flexibility to use Perceiving by taking the following actions:

☐ Allow yourself to have one day per month when you do not schedule or plan the day in advance. See what happens that makes the day worthwhile.

☐ Give yourself time to set aside what you would normally consider a completed piece of work and revisit the work the next day. Note what changes in the work occur to you by allowing the extra time.

☐ Be patient with other ways of doing work, and recognize that progress is being made even if work styles differ.

☐ When solving a problem, list several options in addition to the one that you think is correct. List the advantages and disadvantages of each option and the impacts of each option on people.

☐ When asked for your opinion, give several options and let others decide for themselves.

☐ Realize that rushing to conclusion can actually delay closure if it is the wrong conclusion.

☐ Realize that you are not always right.

☐ Provide opportunities for others to discuss options.

☐ Avoid micromanaging the efforts of others; trust their abilities.

☐ Listen to others and be willing to change your opinion.

☐ View setbacks as opportunities.

☐ When talking to others in person or on the phone, give them your full attention; do not do work at the same time.

Perceivers can increase their flexibility to use Judging by taking the following actions:

☐ Place imaginary deadlines on yourself for generating ideas or gathering information. When you reach the deadline, stop.

☐ Attend a time management seminar.

☐ Schedule extra time beyond what you would normally allow to get to your meetings or appointments.

☐ Identify four to five major things that you need to accomplish in the next year and schedule them. Allow plenty of time to accomplish them, including some contingency time.

☐ Consciously complete some selected tasks ahead of the deadlines.

☐ Periodically commit to challenging goals—and meet them.

☐ Keep a tickler file to remind you to follow up.

☐ Prioritize to-do items and work on the high-priority items first.

☐ Start and stop meetings on time.

☐ Promptly respond to phone calls and requests.

☐ Consider options before a meeting and be ready to state a clear position.

Figure 2-5 *How Behavioral Style Preferences May Affect Leadership Effectiveness*

Leadership Assessment category (see Figure 1-2)	Myers-Briggs preferences likely to reinforce effective leadership	Myers-Briggs preferences likely to be a challenge to effective leadership
1. Be honest, establish trust	*Doing what you say you are going to do and behaving in a manner consistent with your stated values are the keys.* However, we have noticed a trend for **Introverts** to be rated somewhat higher in this category. This may result from the tendency of Introverts to thoroughly think things through before talking about them. As a result, they usually deliver well-thought-out, consistent information.	**Extraverts** tend to think out loud and bounce ideas off of others as part of their thought process. They may have little or no commitment to some of the thoughts they are voicing. They may appear to be inconsistent to those who don't understand their tendency to think out loud. The perceived inconsistency may raise questions about their credibility.
2. Get people involved	**Extraverts** are energized by interaction with others and are likely to engage others in discussions about the project. Extraverts show their enthusiasm and excitement, which attracts others to their cause. **Perceivers'** lack of urgency about getting to closure provides time for others to voice their concerns and ideas, providing a sense of collaboration. Perceivers remain open to options and are eager to explore them with the project team.	**Introverts** may be just as enthused and excited about the project but are less likely to express this excitement and enthusiasm to others. They may not appear approachable. Introverts' energy can be drained by extensive interaction with others, so they may not seek the personal interaction that leads to effective involvement of others. **Judgers** may be so focused on closure and completing work ahead of deadlines that they do not take the time to adequately involve others.
3. Encourage contrary opinion	**Intuitives** are energized by finding new ways to do things. **Thinkers** enjoy debating various views with others and probing others to test their knowledge.	**Sensors** prefer to rely on experience and approaches that have worked in the past so are less likely to be seeking contrary opinions on how to approach project issues or new ways to do the project work. **Feelers'** desire for harmony may reduce their desire to seek out and discuss contrary opinions.
4. Establish a vision	**Extraverts** are likely to show enthusiasm and excitement for their vision. They are likely to have an animated presentation style that will attract others to their vision. **Intuitives** are energized by looking to the future and developing the big picture view of where the project is headed. They are likely to take action on their vision with great confidence that inspires others.	**Introverts** may be just as enthused and excited about their vision but are less likely to express this excitement and enthusiasm to others. **Sensors** are more interested in a step-by-step plan to do the work and are energized by doing project tasks rather than thinking about the big picture. They are more interested in the here and now than in the future.

continued on next page

Figure 2-5   *How Behavioral Style Preferences May Affect Leadership Effectiveness (continued)*

Leadership Assessment category	Myers-Briggs preferences likely to reinforce effective leadership	Myers-Briggs preferences likely to be a challenge to effective leadership
5. Take risks	**Extraverts** freely throw out ideas for discussion as part of their process of thinking out loud, which may convince others that Extraverts are more willing to try new approaches and the related risks. **Intuitives** are energized by finding new ways to do things. **Perceivers** enjoy developing and considering alternatives and do not feel pressure to get to closure, which may convey that they are willing to consider many options.	**Introverts** tend to think things through before talking about them and are likely to have more limited discussions about options they have considered. **Sensors** prefer to rely on experience and approaches that have worked in the past so are less likely to be seen as risk takers. **Judgers** may be so focused on closure and completing work ahead of deadlines that they are not open to options that may involve risks of project delays.
6. Create a positive environment	**Intuitives** have little patience with routines, especially ones they see as ineffective and are eager to get rid of. **Feelers'** preference for decision-making criteria related to effects on others demonstrates genuine consideration for others. Feelers will extend themselves to meet others' needs even at the expense of their own comfort. **Perceivers'** lack of urgency for reaching closure keeps them open for continuing interchanges on new information and ideas at any time.	**Sensors** prefer to rely on experience and approaches that have worked in the past so are less likely to change routines even when the time has come to change them. **Thinkers'** enjoyment of intellectually sparring with others and probing others to see what they know can appear to some to be threatening and to be a personal attack. **Judgers** may be so focused on closure and completing work ahead of deadlines that they discourage others from bringing up ideas and may appear abrupt to others. They may transmit a lack of trust that others will complete their work in a timely way.
7. Challenge limiting beliefs	**Intuitives** are energized by finding new ways to do things and may be energized by helping others find new ways to do their work and to grow.	**Sensors** prefer to rely on experience and approaches that have worked in the past so are less likely to challenge beliefs.
8. Choose your reactions	**Introverts** may not readily express their emotions, even though they may be feeling upset. **Feelers'** desire for harmony and consideration of affects on others may temper their reactions in a tense situation **Perceivers'** lack of urgency for reaching closure may lead to them being calm (perhaps even enthused) when new options are introduced.	**Extraverts** tend to vent their emotions at the time they arise, sometimes without considering the effects on others. **Thinkers** tend to be very direct and may not consider all consequences of their reactions. **Judgers** may be so focused on closure and completing work ahead of deadlines that they may find it difficult to be patient with others.

continued on next page

Figure 2-5 *How Behavioral Style Preferences May Affect Leadership Effectiveness (continued)*

Leadership Assessment category	Myers-Briggs preferences likely to reinforce effective leadership	Myers-Briggs preferences likely to be a challenge to effective leadership
9. Recognize performers	**Extraverts** are energized by interacting with others and are more likely to seek out others to recognize them. **Feelers'** concerns for others may lead to appreciation of and more frequent recognition of good performance.	**Introverts** may appreciate the effort by others but are less likely to seek out the people to express their appreciation. **Thinkers** derive self-satisfaction from accomplishing their project work and may not recognize the need of others for recognition of their personal contributions to the team.
10. Be decisive and competent	**Introverts** tend to think things through before talking about them so that they present well-thought-out ideas to others. **Judgers** place a great deal of importance on being on time and planning and structuring their work to easily meet deadlines.	**Extraverts** tend to think out loud and bounce ideas off of others as part of their thought process. They may have little or no commitment to some of the thoughts they are voicing. What sometimes appear to be random thoughts may detract from others' perceptions of the Extravert's competence. **Perceivers** are energized by being spontaneous and like to go with the flow. They are not energized by developing detailed project work plans. This may cause others to question their competence.
11. Align individual and project goals	**Feelers'** preference for decision-making criteria related to effects on others may lead them to strive to align personal and project goals. **Perceivers'** lack of urgency about getting to closure allows others to voice their concerns and goals, creating a sense of collaboration about personal and project goals.	**Thinkers** derive self-satisfaction from personally accomplishing project work and may not recognize the need to seek out the goals of other individuals and align their project assignments with these goals. **Judgers** may be so focused on closure and completing work ahead of deadlines that they may not consider taking the time to seek out the goals of other individuals and align project assignments with these goals. They may consider any such alignment as far less important than the assignments that get the work done in the shortest amount of time.

continued on next page

Figure 2-5   *How Behavioral Style Preferences May Affect Leadership Effectiveness (continued)*

Leadership Assessment category	Myers-Briggs preferences likely to reinforce effective leadership	Myers-Briggs preferences likely to be a challenge to effective leadership
12. Establish and achieve doable goals	**Introverts** tend to think things through before talking about them so that they present well-thought-out ideas to others that clearly describe their goals. **Sensors** prefer practical, step-by-step approaches and are energized by carrying out tasks involved in a project. **Judgers** place a great deal of importance on structuring their work in a systematic way that will assuredly meet deadlines and achieve the project goals.	**Extraverts** tend to think out loud and bounce ideas off of others as part of their thought process. They may have little or no commitment to some of the thoughts they are voicing. What sometimes appear to be random thoughts may confuse others about the clarity of the Extravert's goals. **Intuitives** find it stressful to attend to details and find considering future possibilities to be more exciting than immediate tasks. **Perceivers** are energized by being spontaneous and like to go with the flow. They are not energized by developing detailed project work plans that include specific milestones and goals. Perceivers tend to be energized by deadline pressure, which may cause others to question if they are going to achieve goals in a timely way.

Figure 2-6 *Style Flexibility Enhancement Action Plan*

Indicate the strength of your preference by placing a mark at the appropriate point on each scale:

Strongly prefer   No strong preference   Strongly prefer

Introversion_____Extraversion

Sensing _____Intuitive

Thinking _____ Feeling

Judging _____ Perceiving

In the "Action List" table below, summarize the action steps that you noted as you read Chapter Two for each preference scale. Considering the relative strengths of your preferences as plotted above, prioritize the action steps in the order that you think they would best assist you in becoming a more effective leader. Establish a schedule for the highest priority step. Move on to the second priority after you have become proficient and comfortable with the top-priority step.

Action List

Style preference	Priority	Action step	Schedule
Extraversion–Introversion			
Sensing–Intuition			
Thinking–Feeling			
Judging–Perceiving			

THREE

Are We All Pulling in the Same Direction?
Defining the Customer's Needs and Expectations

Understanding Needs and Expectations: Precursor to Project Success

The fastest and strongest dogsled team cannot win the race without knowing the direction to the finish line. Similarly, a project cannot succeed unless the project leader and every member of the project team understand what it will take to meet the customer's needs and expectations. You cannot develop a project scope of work, schedule, and budget that meets your customer's needs and expectations if you don't understand these needs and expectations. As obvious as this seems, we have seen many projects run into difficulty because the project team's scope of work does not truly align with the customer's needs and expectations. Remember the story of the Edsel in Chapter Two? The scope of work for Ford's top team of engineers called for them to produce a car that was an advanced mechanical engineering product—which they did; but the Edsel was a failure because the customers did not buy the car. Beta-format video recorders had a similar fate. Sony's engineers met their scope of work that focused on producing a Beta format that produced superior recording quality. However, the manufacturers of VHS-format video recorders were the first to offer two-hour recording times, enough to record most movies, while Beta tapes were, at that time, limited to one hour. The VHS manufacturers understood that the customers were more concerned about the length of recording time than what they judged to be relatively small differences in picture quality. VHS won the approval of the customer and the Beta format faded from the consumer market. To understand the customer's needs and expectations, *Content, Procedural,* and *Relationship* issues all must be understood, not just the *Content* issues on which many project managers want to focus.

Imagine the chaos that would result with a dogsled team if each dog pulled toward a different finish line or in a different direction. It is crucial that the team leader and the team members all understand and share the correct understanding of the customer's needs and expectations. The project leader must have an effective dialogue with the customer to clearly define the customer's needs and expectations before the correct project scope, budget, and schedule can be developed. Once there is a meeting of the minds with the customer, the leader must convert this understanding into a clearly written project scope, schedule, and budget. The leader also must effectively and continuously communicate this understanding to the project team. Chapter Four addresses scoping, scheduling, and budgeting. We devote Chapter Three to understanding the customer's needs and expectations, because this is a vital step that must be accomplished before you develop a scope, schedule, and budget.

Meeting customer needs does not ensure that you will meet customer expectations. Perhaps a real-life example we experienced will clarify this distinction. A few years ago, the City of Edmonds, Washington, was required by the state government to expand and upgrade its wastewater treatment plant on a three-year schedule. Four blocks to the north of the existing, outdated plant was the city's core business district. Immediately south of the plant was a wetlands area. There were expensive condominiums overlooking Puget Sound directly across the street east of the plant. On the west side of the plant was a state highway leading to an adjacent ferry dock used by thousands of motorists every week. There was substantial community interest in looking for other plant sites. The basic *need* of the city was completion of an expanded, upgraded wastewater treatment plant within three years. After extensive discussion between the consultant, city staff, and city council, the sides of the Triangle of Needs contained many *expectations* on the *Procedural* and *Relationship* sides of the triangle (see Chapter One) as well as the basic *Content* need of an expanded plant completed on schedule and within budget:

- Completion of an expanded and upgraded wastewater treatment plant within three years
- An operating plant that meets all state and federal treatment requirements
- Completion of plant construction at costs within the city's financial capabilities
- Operating costs consistent with the city's financial projections and capabilities

- An evaluation process that identifies and objectively considers all available alternative plant sites
- A public involvement process that results in selection of a plant site that is acceptable to the residents and that is environmentally sound

- A process to make information on the project's status during planning, design, and construction available to the residents, city staff, and city council at all times
- A process to immediately address the concerns of any resident during construction and operation of the plant
- A planning process that maintains the city's eligibility for any available state or federal financial assistance

- A project leader and a public involvement leader, both of whom are trusted and respected by all parties
- Continuity of the personnel on the consultant team for the duration of the project
- Assuring the public that their concerns are heard and addressed
- Assuring the public that they have ready access to all project information
- If the plant remained at the existing site, aesthetic design features that establish a positive relationship with nearby businesses, residents, and motorists on the way to the ferry dock
- An evaluation of alternative plant sites that maintains the credibility of the city staff and council with the city residents
- Minimal construction impacts on businesses and residents so as to maintain positive relationships

It was only by careful consideration of the customer's *expectations* that a successful project scope addressing all three sides of the Triangle of Needs was developed.

Once the expectations as well as the basic needs were understood, a scope for the services needed to meet both the needs and expectations was developed, agreed to, and carried out. The ultimate success of the planning phase of the project was evident at the final public hearing on the selection of a plant site. Remember that there was substantial interest in moving the plant away from the center of town when the project planning began. Indeed, residents—including those who lived across the street from the existing plant site—approached the city council at the final hearing with a petition signed by hundreds of city residents. However, rather than calling for moving the plant, the petition called upon the city council to expand the plant at the current site in the heart of town. Meeting the expectation for an inclusive, credible public involvement process resulted in what at the start of the planning phase would have appeared to be an unimaginable result: a demand by the residents that the plant not be moved to a remote site. The plant, expanded and upgraded at its existing site, incorporated aesthetic features identified as desirable by public input during the planning phase that have created a very positive image of the downtown area. These features include a public plaza built over a portion of the plant that offers a view of Puget Sound and that has been used for a variety of public events, including weddings. Taking the time to understand both the needs and expectations of the customers (the city and its residents) was the key to the success of the project.

Here are some questions related to the Triangle of Needs that can be asked of the customer to help determine the customer's needs and expectations:

Issues

- What do you see as the biggest issues on this project?
- What are the biggest concerns to your neighborhood or constituents?
- What are your budget constraints?
- What are your time constraints, and what is driving those constraints?
- Do you have any example outputs from previous, similar projects that you can share with us? What did you like or dislike about them?
- If this project were an unqualified success, what would it look like?

Issues

- What are the lines of communication with your organization or group for this project?
- How and when do you want to get progress updates?
- What is the best way for us to work with you in presenting deliverables/products for review so that we can get all of your comments in a timely way?
- Who in your organization will be involved in giving input and making decisions about this project?
- What is the best way for us to deal with any changes in project scope that may occur due to changing conditions or identifying additional needs or aspects as the project proceeds?

Issues

- Who are the stakeholders involved in this project?
- What are some of the stakeholder expectations for this project?
- What are the relationships between the stakeholders?
- Are there any conflicting opinions about this project among the stakeholders?
- From whom do we need to get buy-in for the project results to be accepted?
- What level of your involvement works for you on this project?
- What has worked well for you (or not so well) in working with other project teams on other projects?

Personal Styles and Their Effects on Defining Needs and Expectations

The personal style preferences of the customer and team leader can contribute to some pitfalls during the process of defining the customer's needs and expectations (Brock, 1994). These pitfalls and some suggestions on how to avoid them are presented on page 75.

Style/person	Potential pitfalls	Avoiding the pitfalls
Extraverted customer	• May throw out thoughts about needs and expectations for which he/she has little or no commitment	• Ask questions to see if the thought is important or just part of the customer's thought process • Submit a draft summary of your perception of the critical needs and expectations you heard the customer express before attempting to prepare a project scope; meet with the customer to discuss your draft summary
Extraverted project leader	• May form and express ideas about the customer's needs before fully understanding them • May erroneously assume that silence means the customer has accepted that needs and expectations have been agreed to	• Let the customer speak first, be patient, let him/her finish • Ask questions until you understand the customer's needs and expectations • Listen carefully before speaking • Check your perceptions by asking questions, such as, "If I'm hearing you correctly, it appears your major concerns are . . . is that correct?"
Customer and project leader are both Extraverts	• Both may be talking without hearing and exploring the other's views • Jumping from one topic to another without reaching a conclusion • Treating all ideas as relevant	• See suggestions above for Extraverted customer and Extraverted project leader • Summarize the conclusions reached on one topic before moving on to another • Ask the customer to prioritize the initial list of needs and expectations; this process can result in identifying which are critical and which were just fun to talk about
Introverted customer	• May not readily express all needs and expectations • Will want time to think about needs and expectations before agreeing on them	• Ask questions to draw out the customer's thought process • Submit your preliminary perception of needs and expectations prior to meeting to start the customer's thought process
Introverted project leader	• May let the customer talk himself/herself into a premature definition of needs and expectations	• Ask open-ended questions (i.e., questions starting with what, when, how, where, or why) • Share your thoughts even if you think they are too insignificant to raise
Customer and project leader are both Introverts	• Presenting conclusions without sharing thought process can create tension when the conclusions of leader and customer differ • Leaving unsaid his/her concerns that need to be explored • Underestimating time to reach agreement because both need time to think things through	• See suggestions above for Introverted customer and Introverted project leader • Describe your thought process before presenting your conclusions • Ask questions until you understand the customer's thought process before discussing differences in conclusions • Ask the customer how much time he/she would like to review a draft list of needs and expectations, and abide by his/her schedule

continued on next page

Style/person	Potential pitfalls	Avoiding the pitfalls
Sensing customer	• May want to talk about detailed, step-by-step approach to doing the project work before fully defining the bigger picture of needs and expectations	• Ask questions about how the details raised by the customer relate to underlying needs and expectations • Ask what a successful end result will look like to assist the customer in thinking of the bigger picture
Sensing project leader	• May perceive the customer as rambling because he/she is not talking about details of how the work will be done	• Set aside your desire to talk about the detailed steps involved in doing the work • Realize that what you perceive as rambling may be an important part of the customer's process of identifying needs and expectations • Ask open-ended questions about the customer's needs and expectations, and check your perceptions of his/her answers
Customer and project leader are both Sensors	• Failure to see the interactions between various needs and expectations • Focusing on short-term needs	• See suggestions above for Sensing customer and Sensing project leader • Ask how a specific expectation relates to achievement of a project need
Intuitive customer	• May have many thoughts about needs and expectations but some may be incompatible and/or unrealistic, such as achieving top quality but at the lowest cost	• Based upon a discussion with the customer, prepare a prioritized list of your perception of his/her needs and meet with the customer to reach agreement on these priorities (the prioritization process will focus the customer on defining his/her needs)
Intuitive project leader	• May use ambiguous language or jargon that confuses the customer • May tend to jump from one topic to another when asking questions about needs and expectations	• Put your understanding of needs and expectations into analogies and language that the customer understands • Keep questions simple, with only one major thought per question
Customer and project leader are both Intuitives	• May be short on information that adequately defines needs, which may lead to an approach that solves the wrong problem	• See suggestions above for Intuitive customer and Intuitive project leader • Explore realistic "what if" scenarios for the project to assist the customer in defining his/her needs and expectations
Thinking customer	• May be too focused on the what and how aspects of the project	• Ask questions about who will be impacted by the project and what those people's needs and expectations may be • Ask questions about who will need to buy in to the project results and how that buy-in can best be obtained

continued on next page

Style/person	Potential pitfalls	Avoiding the pitfalls
Thinking project leader	• May define a successful project as solely one that has a technically superior end result, delivered on time within budget • May alter the customer's needs to match the project leader's perception of the right solution	• Recognize that how people feel about the project end result is important in determining its success (remember the Edsel!) • Listen carefully to the customer for intent and feeling about expectations as well as content that defines needs • Recognize that your superior technical knowledge can lead to an approach that does not truly match what the customer needs; ask questions and listen to define the customer's needs and expectations, then think about the approach with an open mind
Customer and project leader are both Thinkers	• May fail to consider impacts on people that will be involved in the project • May identify logical needs but overlook the relationship aspects that are needed for a successful project	• See suggestions above for Thinking customer and Thinking project leader • Get input from others that have a Feeling preference in your organization as well as the customer's
Feeling customer	• May be so focused on the *Relationship* needs of the project that some of the *Content* and *Procedural* needs may not be adequately defined	• Draw the customer into a discussion of how the *Relationship* needs can best be met by the project end result and what procedures might best be used to get to that end result
Feeling project leader	• May shy away from expressing lack of understanding or disagreement about project needs and expectations	• Remember that discussing your questions or even disagreement can lead to a better understanding of the customer's needs and expectations; understanding them now can avoid conflicts later
Customer and project leader are both Feelers	• May be overly influenced by the views of a few trusted people • May primarily identify immediate needs of people • May identify needs that are inconsistent with previous decisions or policies	• See suggestions above for Feeling customer and Feeling project leader • Solicit a diversity of views on project needs from others in your organization and the customer's • Do a reality check on the compatibility of identified needs with previous decisions and policies before drafting a scope
Judging customer	• Desire for closing the discussion may cause some needs to be overlooked or inadequately discussed	• Continue to ask questions until you are sure that all needs and expectations have been identified and adequately explored
Judging project leader	• May not want to take the time to discuss a variety of opinions about needs	• Identify all who should have input on the project needs and include them in the discussion

continued on next page

Style/person	Potential pitfalls	Avoiding the pitfalls
Customer and project leader are both Judgers	• Premature closing of discussion • May ignore new information that affects definition of needs and expectations	• See suggestions on page 77 for Judging customer and Judging project leader • Both parties need to be open to new information that may affect definition of project needs and expectations
Perceiving customer	• May want to keep adding thoughts about needs and expectations, even though agreement seemed to have been reached	• Document the understanding of needs and expectations in writing • Explain that the mutual understanding of needs and expectations that you have reached will be the basis for preparing a scope of work; state that although you will be happy to discuss any added thoughts, it will affect the schedule for preparing the scope
Perceiving project leader	• May not assist the customer in coming to closure on identifying needs and expectations	• Pause, summarize the needs and expectations identified, and ask if the customer has any other thoughts, while avoiding interjecting other ideas of your own about their needs
Customer and project leader are both Perceivers	• May lose focus and get lost in a sea of possibilities • May develop more needs and expectations than needed	• See suggestions above for Perceiving customer and Perceiving project leader • Be sure that needs are all detailed and quantified in the project scope

Converting Needs and Expectations into a Plan

This chapter is devoted to understanding the customer's needs and expectations because that understanding is vital to project success. This understanding establishes your direction. It focuses the energy of you, your team, and your customer, and keeps all of you pointed in the same direction. It creates commitment. It gives you a way of knowing when you are done—just like the finish line does for the dogsled team. Once the customer's needs and expectations are understood, this understanding must be converted into a project plan that includes a detailed description (scope of work) of how the needs and expectations will be met, as well as the schedule and budget for doing so. Chapter Four is devoted to this planning effort.

Exercises

1. Think of an ongoing or past situation with a customer where you have experienced some challenges in fully understanding the customer's needs and expectations. These challenges may have surfaced as scope changes as the project proceeded, or resulted in an unhappy customer who perceived the project as falling short of meeting needs and expectations. Referring to the information on style preferences in Chapters Two and Three, use the following series of questions to identify where differences or similarities in style preferences may have contributed to that misunderstanding:

Extraversion–Introversion

Your preference	Your customer's preference
☐ Extraversion	☐ Extraversion
☐ Introversion	☐ Introversion

- How did/does your difference or similarity in styles contribute to challenges in fully understanding your customer's needs and expectations?
- What action can you take to address these challenges now or in the future?

Sensing–Intuition

Your preference	Your customer's preference
☐ Sensing	☐ Sensing
☐ Intuition	☐ Intuition

- How did/does your difference or similarity in styles contribute to challenges in fully understanding your customer's needs and expectations?
- What action can you take to address these challenges now or in the future?

Thinking–Feeling

Your preference	Your customer's preference
☐ Thinking	☐ Thinking
☐ Feeling	☐ Feeling

- How did/does your difference or similarity in styles contribute to challenges in fully understanding your customer's needs and expectations?
- What action can you take to address these challenges now or in the future?

Judging–Perceiving

Your preference	Your customer's preference
☐ Judging	☐ Judging
☐ Perceiving	☐ Perceiving

- How did/does your difference or similarity in styles contribute to challenges in fully understanding your customer's needs and expectations?
- What action can you take to address these challenges now or in the future?

Figure 3-1 appears on page 81

2. Use Figure 3-1 at the start of your project to establish each individual customer's needs, expectations, project objectives, and key scope issues. Alternatively, complete Figure 3-1 for a recently completed project as "20:20" hindsight. Compare it to what you actually did on the project and highlight what could have been done differently if the customer's needs had been more clearly defined at the start of the project.

3. Develop your own list of questions to ask your customers for each side of the Triangle of Needs for your types of projects. Use the list as a checklist on your next project.

Figure 3-1   *Project Objectives Worksheet*

Customer Name		Style Preference	E I T F S N J P
Project Leader		Style Preference	E I T F S N J P
Customer's Needs and Expectations			
Content			
Procedural			
Relationship			
Project objectives			
Critical issues to keep in mind when developing scope			
Potential areas of vagueness, confusion, or lack of closure			

E = Extraversion; I = Introversion; S = Sensing; N = Intuition; T = Thinking; F = Feeling; J = Judging; P = Perceiving

What Route Are We Taking?

Planning the Project

The Value of Planning

When we ask project teams about the value of planning, team members usually list the following benefits:

- ▶ Provides better understanding of what will happen during the project
- ▶ Uncovers scope, budget, and schedule problems up front
- ▶ Allows better anticipation of resources
- ▶ Provides better anticipation of problems and solutions
- ▶ Increases understanding of the project
- ▶ Improves coordination and communication
- ▶ Builds commitment
- ▶ Provides basis for monitoring and control
- ▶ Increases the commitment of team members who participate

Why Don't You *Always* Plan Your Projects?

When we ask project teams and managers if they always plan their projects before beginning work, we find a substantial majority confessing that they do not. Here is a list of their reasons. Check those that you may have used.

- ☐ I'd rather jump right into the "real" work of the project.
- ☐ I don't have time to do a plan.
- ☐ I don't have the budget to do a plan.
- ☐ Even if I do one, I won't have time to keep it up-to-date.
- ☐ The customer doesn't want to pay for a plan.

☐ I don't understand the project well enough yet to do a plan.
☐ I need to get the team going to keep them busy—I'll do a plan later.
☐ The plan in the proposal is good enough.
☐ This project is going to have changes, so why bother?
☐ The project is too small—it will be over before the plan is complete.

Imagine our dogsled team trying to start a race without knowing the location of the starting line, the finish line, or the route and distance in between. They would no doubt get off track and probably run out of provisions on a long race, retrace their steps until they found the right path, and have no chance of winning the race. As popular as the above excuses may be, beginning a project without a plan is assuredly going to lead to problems. A project *leader* always sees that a project plan is completed before the project work starts. Without a project plan, you'll end up redoing work, wasting time, searching for a path to your objective, and making unnecessary expenditures. Without a plan, project leadership is impossible, and project management becomes a random event with an uncertain outcome.

There are few guarantees that we can offer, but here is one:

> **An hour spent in planning the project**
> **will save more than an hour in doing the project work.**

The return is more likely to be ten to one. If we were to say to you that, if you will give us one dollar, we will give you ten dollars with no strings attached, would you take us up on it? The next time that you sense you are about to use an excuse for not planning the project, remember that you are about to lose the chance to invest one dollar to get ten dollars.

Balancing the Triangle of Needs When You Plan

As in every aspect of a project, a project leader keeps all sides of the Triangle of Needs in mind and in balance as a project plan is developed.

Content needs are met when planning the project by:
▶ Defining a project scope of work that fulfills the customer's needs and expectations
▶ Breaking down the scope into well-defined, manageable tasks
▶ Identifying a schedule with milestones
▶ Identifying the resources and budget needed to accomplish the project
▶ Defining any training needed to accomplish the work

Procedural needs are met when planning the project by:
▶ Defining communication procedures
▶ Defining progress reporting procedures
▶ Conducting a planning workshop to define goals, constraints, and roles

- Developing a flow chart that describes how the project work will proceed and the relationships between project tasks
- Identifying potential issues that may arise and how they will be addressed
- Defining a conflict resolution process
- Defining a process for dealing with scope changes
- Preparing a written project plan

Relationship needs are met when planning the project by:
- Defining customer needs and expectations
- Involving the team in planning the project
- Identifying and addressing team concerns about the project
- Conducting a project planning workshop where all team members participate
- Devoting part of the planning workshop to building team spirit, establishing trust, and defining accountability
- Aligning project assignments with individual development goals
- Developing a plan and responsibilities for frequent communication with the customer to determine how the customer feels about the progress of the work

Planning the Team

In Chapter Two, we used an analogy of right-handedness and left-handedness to illustrate the use of different personal style preferences. While we may prefer to use our right or left hand, every able-bodied person uses both hands. Consider what it would be like if you chose to use only your preferred hand. Using both hands increases the range of things you can do and increases your overall ability. There is a risk in planning your team that you may become too "one-handed."

There is a tendency to select individuals who have style preferences similar to your own. This can lead to team blind spots. For example, a team that has a majority of Sensors, Thinkers, and Judgers may be so focused on rational analysis of detailed information in their drive to get to closure that they may have difficulty in dealing with changing circumstances. A team with a preponderance of Intuitives, Thinkers, and Perceivers may be a dynamic, risk-taking team that may overlook early signs of problems that may doom their project to failure. Building a team that is too one-handed can lead to the blind spots the Edsel design team encountered, as discussed in Chapter Two.

Be aware of your team members' style preferences and the benefits of including diverse style preferences. For example, consider the benefits you may get from a balance of Sensing–Intuitive and Thinking–Feeling preferences on your team. Sensors gather the facts, look at the details, are specific and clear. Intuitives look at the facts and how they fit into the bigger picture, consider various alternatives, and develop possibilities. Thinkers analyze the alternatives objectively and consider the cause and effect of potential actions and the consequences of alternatives. Feelers look at how the decision will affect others. Having a mix of Perceivers and Judgers is also helpful, in that Perceivers may keep the Judgers from rushing to a premature conclusion, and Judgers will help the Perceivers make decisions and move forward.

Of course, keep in mind that personality and related behavior involves many factors other than personal style preferences. The mix of team members still must be able to work together productively, and the chemistry among team members also must be considered. If the team is very small, diversity potential is limited, and you should consider including some individuals with style preferences that are different from those of the majority of team members in periodic reviews of the team's work.

There are other factors to consider in planning your team. Availability and capabilities need to be balanced. All project leaders are seeking the same, most productive people in an organization to be on their teams. It does your team no good to have a superstar on your team if that superstar's time is fully committed to other projects. A frank discussion of availability should be held with each potential team member.

The desire for diversity of styles and skills needs to be balanced with the number of people on the team. The number of communication links and the related opportunity for miscommunication go up dramatically as the number of team members increase. There are twelve communication links on a four-person team and ninety communication links on a ten-person team. The chances for miscommunication go up as the number of communication links increases.

Also, individual productivity may decrease when team size increases. German psychologist Max Ringlemann compared the results of individual and group performance on a rope-pulling task (Robbins, 2002). He expected that three people pulling on the rope should exert three times as much pull on the rope as one person, and eight people should exert eight times as much pull. The results were quite different. Groups of three people exerted a force only 2.5 times the average individual performance. Groups of eight collectively achieved less than four times the individual rate. Other researchers have confirmed Ringlemann's research in studies involving group tasks. Even though the total productivity of a group increases with size, the individual productivity of each group member tends to decline. The effect has been called "social loafing." One explanation is the dispersion of responsibility. Because the results of the group cannot be attributed to any single person, the relationship between an individual's input and the group's output is masked. As a result, individuals may tend to rely more on the group's efforts than their individual efforts. Leaders counteract this effect by providing means to identify and measure individual efforts.

When you don't have the luxury of picking your team, awareness of the styles you have will enable you to capitalize on the strengths and compensate for the weaknesses. Consider your team: using the chart below, what are the strengths and weaknesses of the team members?

Name	Role	Style preference	Style-related strengths	Style-related weaknesses
(You)				
Team member				
Team member				
Team member				
Team member				
Team member				
Team member				

Actions I can take to capitalize on these strengths and compensate for the weaknesses:

Developing the Project Scope of Work

After understanding the customer's needs and expectations, the project objectives should be clearly defined in a concise statement that will serve as the basis for the scope of work. To be effective, the objectives should have several key characteristics:

- ▶ *Be specific.* Anyone with basic knowledge of the project subject should be able to read the project objectives and understand them.
- ▶ *Be measurable.* All project objectives have a measurable schedule, scope, and budget.
- ▶ *Be realistic.* Objectives that are challenges are desirable because they motivate the team, but they need to be achievable. The availability of personnel, other resources, and timing must be considered in setting the objectives to avoid impossible deadlines, inadequate funding, or inadequate staff.
- ▶ *Be in agreement.* You, the customer, and the team all must agree on the project objectives.
- ▶ *Be assignable.* The objectives must be capable of being divided into tasks for which an individual will be responsible. This allows for accountability and project control.

The Project Objectives Worksheet in Figure 3-1 can be used to assess customer needs, expectations, and issues, and should be used in defining the project objectives and in developing the scope of work.

Once the objectives are clearly established, you need to keep everyone focused on them. Distribute a concise written statement of the objectives to the team and the customer and constantly remind everyone of the objectives. Constantly remind the team members that their efforts are to be focused on these objectives. If they are doing something that does not advance them toward the objectives, it's time to reassess and align their activities with the objectives. Lead dogs always have their eyes on the finish line.

With the objectives defined, the scope of work must be developed. The following are some considerations related to your customer's style preferences to keep in mind.

▶ *Extraverted Customer.* Relying on the Extraverted customer's evolving discussions of what they want from the project is a major pitfall to avoid. If you base your project scope solely on what your Extraverted customer says he/she needs, you may not produce a scope that meets his/her true needs. Also, if everything the Extraverted customer says is treated as gospel and incorporated in the project scope, the customer may have severe sticker shock when the team later submits the project budget. Submit a draft scope for the project to the customer before preparing the budget to be sure that you understood what is truly important to the Extraverted customer.

▶ *Introverted Customer.* Recognize that the Introverted customer will want time to reflect upon a proposed scope. Submit a draft scope prior to meeting to discuss the scope so that the Introvert's thought process can begin prior to meeting. Arrive prepared for the scoping meeting, because Introverts do not like to shoot from the hip.

▶ *Sensing Customer.* A Sensing customer prefers project scopes and goals that are straightforward. Sensors like scopes that are clearly doable. Project goals that may be inspirational to an Intuitive may not seem real to a Sensor. Sensors like to get their arms around the scope and see how the work can be done. They will be looking to define the project with a detailed scope that spells out how the work will be done in a step-by-step manner. Help the Sensor see how these steps fit into the bigger picture of what is to be accomplished by the project and what the end result can look like. For example, if the project involves a feasibility study, ask the customer for a sample of a similar study that they've liked. If they don't have one, show them a project feasibility study you have done as an example. By helping them visualize the final study by showing them examples, you can clarify the project scope in the context of the bigger picture.

▶ *Intuitive Customer.* Intuitive customers want project scopes that inspire and challenge them. Intuitives are usually happy with broader definitions of the project than a Sensor and don't have as much interest in the detailed steps involved. Intuitives often think that the detailed steps will be obvious, and Intuitives do not like to talk about the obvious. Take the initiative to see that the Intuitive gives enough attention to the details of the project so that the scope is grounded in reality. The Intuitive customer may use language that can cause problems when describing objectives, due to their inherent ambiguity: "optimum," "maximum (or minimum) extent possible," "approximately," "at least," "include but not necessarily limited to," and "nearly." If there is a term that is open to varying interpretations, it doesn't belong in the scope.

▶ *Thinking Customer.* A Thinking customer is looking for a scope that is the result of a logical thought process and that includes the "what and how" aspects of the project. Thinking customers are looking for a rational path for the project work to proceed to the end results and benefits that the project will produce. Prepare a project scope that includes a logical sequence of clearly defined tasks and related schedule.

▶ *Feeling Customer.* A Feeling customer is more concerned with the "who" (who will be affected by this project scope) rather than the "what and how" aspects. A Feeler will be concerned about whether people will be better off as a result of the project and how the project will affect the quality of life. The discussion of the project scope needs to include identification of the opportunities for stakeholder input during the project work to assure the Feeling customer that the impacts on people will be adequately considered.

▶ *Judging Customer.* A Judging customer will want the project objectives to be explicitly defined in the scope so that everyone agrees to them. When agreement is reached, the Judger considers the process completed, unless some significant change in circumstances occurs. Because of a Judger's desire for closure, be sure that all alternatives that may affect the scope have been considered before closing the discussion and finalizing the scope. Discuss the time required to achieve the project objectives so that the Judging customer realizes that all of the objectives may not be reached at the same time.

▶ *Perceiving Customer.* For a Perceiving customer, the process of agreeing on the project objectives is an ongoing process with different levels of agreement. To the Perceiver, the project objectives are guidelines open to reevaluation and additional information. When dealing with a Perceiving customer, it is especially important that the project scope be very carefully defined in writing so that he/she can clearly identify changes he/she may later create by altering the objectives and can understand the impact on schedule and budget. Discuss the procedure by which changes in project scope will be addressed as the project proceeds.

Figure 4-1 appears on page 98 A Project Scoping Checklist is provided in Figure 4-1. Use it during the scoping process to assist you in addressing the critical items.

Preparing for Scope Changes

Because no project goes exactly according to plan, it is likely that changes in scope will occur during project execution. These changes may have an effect on the project schedule and budget. It is prudent to discuss these facts with the customer and to establish a procedure that will be used to address such changes before work starts on the project. It is much more difficult to establish such a procedure at the time the first scope change occurs, when time and patience to do so may be limited. The procedure should identify what information is to be supplied in a request for a scope change and who has the authority to approve the scope change. It will be useful to develop a form for requesting scope changes that includes information such as the following:

- ► Project name
- ► Who is requesting the change
- ► Date of the change request
- ► Tracking number of the change request
- ► Reason for the change
- ► Description of the change
- ► Justification for the change
- ► Effect on the schedule (if any), with supporting detail
- ► Estimate of additional cost (if any), with supporting detail
- ► Signature line for organization submitting the request, with date of submittal
- ► Signature line for approval or rejection for the customer with date of the action

Figure 4-2 appears on page 100 Figure 4-2 is an example scope change request form. Provide information to the project team so that they know the procedure for documenting and requesting changes in scope.

Breaking Down the Scope into Well-Defined, Manageable Tasks

In order to do the project, the project leader develops an outline that divides the objective into tasks, each with its own objective, hours, schedule, budget, defined inputs and outputs, and a responsible individual. The task descriptions tell who will do what, when they will do it, how much they will spend to do it, and how progress will be measured.

Your goal should be to use the *minimum* number of tasks necessary to describe the work. A large number of tasks does not lead to clearer definition and more control. Instead, it may lead to a loss of control, because the people working on the project will not be able to track how their time is divided among a multitude of tasks. Also, you won't have time to effectively track numerous minor tasks. The task outline should be as simple as possible so that it can be easily tracked and updated. If you are an Intuitive who likes to focus on the big picture, be sensitive to the fact that you may fail to break down the project into tasks small enough so that each task is manageable. Sensors, on the other hand, may tend to break down the project into too many tasks, each described in more detail than necessary. A balance between the Intuitive and Sensing preferences is needed.

In defining the tasks, use the *same* list of tasks from the scope of services in the project contract, in the project schedule, and in the project budget. By being consistent, it will be easier to track progress and relate it to your original plan and contract. It also will be easier to document any changes in scope as the project proceeds.

The following items should be included in a task definition:

▶ *Task Summary.* A brief (three to four sentences) description of the purpose and objective of the task.
▶ *Schedule.* Start and finish date for the task.
▶ *Budget.* List of the individuals who will perform the task, the number of hours budgeted for each individual, and the total dollar amount for labor and expenses for the task.
▶ *Schedule.* Subtasks comprising the task and the completion dates/milestones for each of these subtasks.
▶ *Output.* Description of what the task will produce (e.g., a report, design memo, drawings, specifications, prototype)
▶ *Input.* List of the information required from other tasks and when that information is needed.

Planning Workshop

A planning workshop at the start of any project, regardless of size, is vital in satisfying *Content, Procedural,* and *Relationship* needs and to establish a common understanding of the project. The benefits of a planning workshop include the following:

▶ Tasks are defined from all points of view and using the strengths of differing style preferences

- ▶ All team members are focused on planning ahead
- ▶ Communication procedures are set up internally and externally
- ▶ Decision-making processes are defined
- ▶ Early review by team reduces errors

- ▶ Team members get to know one another and discover goals, concerns, strengths, and weaknesses
- ▶ Team member roles are discussed and agreed upon
- ▶ Synergism occurs—experienced team members can help inexperienced team members
- ▶ Conflict potential is reduced
- ▶ Team identity and cooperation are developed

Don't assume that the project will be completed free of problems and changes. The chances of success increase greatly when problems and related solutions are anticipated. An essential part of the planning workshop is the brainstorming of major potential problems and what actions might best address these problems. An effective project leader recognizes that differences in personal styles can be an asset or an impediment to the brainstorming process. Chapter Two describes an effective method of brainstorming that capitalizes on the strengths of both Extraverts and Introverts.

As a result of brainstorming, the team might, for example, identify the fact that obtaining a permit for a project could cause a major delay. The action steps could include an early definition of permit submittal requirements, early collection of the needed information, and assignment of an individual to monitor the progress of the application. An action might also involve adding someone to your team who has specialized skills to resolve the problem. Involvement of others may be increased or decreased as problems are considered. By considering alternative pathways to the project objective, the team will be better equipped to cope with problems if or when they arise.

The Extraverted project leader is eager and enthusiastic about describing his/her ideas for the project to the team. The leader's enthusiastic presentation of the project vision and objectives is important. However, the leader must remember that the planning workshop is also an opportunity to benefit from *listening* to team members' suggestions for the project. If the planning workshop turns into a one-way presentation by the Extraverted project leader, the team's commitment to the project will likely decrease because the team members will not have the feeling of involvement that comes from a truly participative planning effort. While the Extraverted leader may look forward to and enjoy the verbal interchange with the team during the planning workshop, the task of preparing the written project plan will probably be considerably less appealing. Recognizing this tendency, the Extraverted leader may find it useful to designate, in advance of the planning workshop, one of the team members to prepare the initial draft of the project work plan.

The Introverted project leader needs to recognize that he/she needs to visibly express enthusiasm and excitement about the project. Such an expression will require an Introvert to make a conscious effort to increase his/her energy level while describing the project visions and objectives. The Introverted leader has a tendency to internalize a great deal of information as well as the thought processes that lead to the conclusions on how the project should proceed. The team members may view this behavior as an attempt to control them through withholding information. The Introverted project leader needs to share all relevant information with the team and describe his/her thought processes in the planning workshop before discussing specific ideas with the team.

A project leader is also sensitive to the differing styles of the team members during the workshop, calling on the Introverted team members to solicit their ideas while controlling the amount of speaking time used by the Extraverted team members. The project leader recognizes the potential limitations of personal preferences (both his/her own and those of the team members) on the planning process and will call upon team members with varying preferences to get a balanced approach to the project. Sensors can help the Intuitives pay enough attention to details to make the plan workable. Intuitives can help the Sensors develop details that are consistent with the overall objectives. Thinkers can help the Feelers see logical methods and procedures for carrying out the project work. Feelers can help Thinkers remember that they need to consider the impacts on people as the work proceeds and that the project plan honors the customer's and the team's values. Judgers can help Perceivers realize when enough alternative approaches to the project have been considered. Perceivers can help Judgers from prematurely deciding that "the plan" is ready to go.

Figure 4-3 appears on page 101 Figure 4-3 presents a checklist of topics for the project planning workshop. Note that the checklist suggests devoting part of the workshop to identifying and discussing the Myers-Briggs style preferences of the team members. This requires that each team member complete a Myers-Briggs questionnaire in advance of the workshop and that the workshop include a facilitator qualified (if the team leader is not qualified to do so) to lead a discussion of the team preferences and how they may affect the conduct of the project work as well as the relationships among team members. Although inclusion of this topic requires more effort than often is associated with a typical project kickoff meeting, the potential dividends are great in terms of better working relationships among team members. When the team members understand how their style preferences affect their different approaches to planning and conducting the work, misunderstandings and tensions among team members will be greatly reduced.

Project Work Plan

After the workshop, the input must be incorporated into the project work plan including a written description of the tasks and the essential task elements listed earlier. Circulate the written material to the team for comment and have back-and-forth discussions until there is agreement. This may require an intense effort on your part to accomplish this in a reasonable time, with a lot of one-on-one discussions after the workshop. The Extraverted leader will enjoy the prospect of the one-on-one discussions of the draft plan with the team members, but he/she must recognize that the results of these discussions must be promptly translated into the written plan and closure on the final project plan must be reached in an expeditious manner. The Introverted leader may prefer to refine the draft plan via e-mails and written memos but needs to overcome this tendency by recognizing the value of face-to-face discussions with team members. Leaders with a preference for Judging need to be patient and ensure that they have given all team members ample opportunity for input. On the other hand, leaders with a preference for Perceiving need to recognize when the point of diminishing return from the team's input has been reached and that it is then time to come to closure on a plan.

The rewards in terms of fewer problems as the project proceeds justify the effort of a planning workshop and subsequent discussions with team members. When tasks are understood and are within the capabilities of the team members, there will be a high degree of confidence and commitment. By opening the project plan to discussion, you foster the involvement that leads to team commitment. It is still your responsibility as a project leader to see that all the pieces are pulled together into a coherent, rational plan. Judgers need to be especially sensitive that the plan needs to be a dynamic tool to allow the team to cope with changes in an orderly manner. As much as Judgers would like to operate in a world of certainty, some things will go wrong or conditions will change as the project proceeds. There is no such thing as a perfect plan that anticipates every problem or condition. Recognize that problems and changes will arise, and have a plan to minimize impacts. Issue revisions to the plan as changes occur. Be sure that everyone on the team gets every revision promptly.

Figure 4-4 appears on page 102

Figure 4-4 is a checklist of items to consider for the project work plan.

Don't assume that your customer will not pay for the effort needed to prepare a project work plan. When the work plan is identified as a deliverable item in the project scope and the content and value of the plan are understood, the great majority of customers will agree to include the preparation of the plan in the project budget.

PLANNING QUIZ		
	Yes	No
Does your current project have a detailed work plan?		
Do all of your team members have a copy of the work plan?		
Is the work plan up-to-date, with all changes included?		

If the answer to any of the above is "No," put the plan on the agenda for the next team meeting and see that appropriate positive steps are taken.

Project Scheduling

When the objectives and tasks are defined, your team needs to develop a project schedule. As you work with others on your team to define the schedule, remember that tolerance for risk varies substantially from person to person. Some will naturally give you optimistic schedules while others will be very cautious, adding in their own contingency time. Intuitives often struggle to come up with realistic time requirements for tasks and need to call upon their Sensing preference, or others on the team who have a Sensing preference, for a reality check. Disagreements about planned task durations often relate to differences in personal styles and tolerance for risks rather than about the time needed to do the work.

As you assess the reasonableness of the time for each task, you must evaluate it in terms of the capabilities of the people who will actually do the work. A common pitfall is to evaluate the time required as if *you* were going to personally do the work on the task. The people doing the work may be either more or less expert than you, so adjust your estimate of the time required in light of their abilities and your past experiences with similar tasks. If the task is something that you've been involved with many times in the past, you have a good basis for evaluation. For example, someone projects a time of ten hours to review a ten-page chapter in your team's report. You know from experience that it should take one to two hours, depending on the complexity of the section and the skill of the author. You have good reason to question the estimate.

Figure 4-5 appears on page 103

Be sure to include contingency time. All projects deviate from plan at some point. If there is no contingency time, the first missed task deadline creates an impossible chain of other task deadlines and the team becomes discouraged. Figure 4-5 presents a checklist of scheduling items. Use it to see if your current project schedule needs adjustment.

Project Budgeting

You can't achieve your project objective and schedule without an adequate budget. Don't attempt to prepare the budget until you have completed the project scope and schedule. There is no way to accurately estimate the budget without both the scope and schedule because they determine the resources needed. We've seen more than one case where preoccupation with the financial aspects of a project caused the budgeting step to be taken out of this logical sequence. It doesn't work.

There are two basic approaches to preparing a budget: "top-down" and "bottom-up." In the top-down approach (which appeals more to Intuitives and makes Sensors uncomfortable), the overall project cost is projected based on some experiential factor, such as dollars per square foot of building, the cost per pound of fabricated steel, or the cost per line of programming. Whatever business you are in, there should be some historical costs of projects that can be used as a rule of thumb, top-down estimate. Historical cost estimates should be used with caution, however, because no two projects are the same. A top-down estimate is not directly related to the level of effort needed for the project. The task budgets are then developed by working backward from the top-down total. Although this may be a useful reality check on a budget, it is not a good way to prepare a project budget. It does not provide good planning and control information, because the resulting task budgets may not have any relation to the required effort.

In the bottom-up approach (which appeals to Sensors but tries the patience of Intuitives), the hours of effort for each task are estimated and multiplied by the labor rates of those who will do the work to determine the labor costs. Material costs and other expenses also are estimated for each task, as are the appropriate contingencies for each task. This approach has the advantages of forcing the Intuitives on the project team to consider the requirements of each task, providing a meaningful basis for project control, and forming a good basis for negotiations of project costs with your customer. The detailed tasks, scopes, and budgets will give your customer a clear picture of the cost and time implications of various aspects of the project. If the customer wants to alter the scope, he/she has a basis by which to judge the effects. If the customer has champagne tastes and a beer budget, the detailed task scopes and budgets often result in appropriate increases in the project budget or a decrease in project scope.

Compare a bottom-up estimate with a top-down estimate to determine reasonableness. If the top-down estimate is considerably higher, look for costs you may have missed that were not directly related to the tasks (don't forget project administration costs!) you've identified, or for some unjustified simplifying assumptions that have been used. If the bottom-up estimate is higher, be sure that you haven't included more detail than necessary or used multiple contingencies on the same work items.

As in schedule development, involve the team members responsible for doing the work in estimating the cost, and use their varying personal style preferences to create the strongest possible plan. As was the case in estimating time, there is no way to be certain of the cost of each task. It is prudent to include a contingency allowance on a task-by-task basis because the risks vary from task to task. For example, the contingency time and budget associated with writing a section of the project specifications can be less than that involved in obtaining a permit from a government agency. Be careful not to add contingency upon contingency because multiple contingencies become unreasonable. Apply appropriate contingencies to each task based upon its uncertainty, rather than apply a blanket contingency to all tasks. Use the total of the task contingencies as the project contingency. Transfer funds from the project contingency back to individual tasks as the need arises.

Don't plan costs in any greater detail than you will be receiving from the project cost information. It's pointless to budget costs for subtasks for which no separate cost information will be reported or to budget weekly costs if only monthly cost reports will be available.

There are several points of caution to consider in your budgeting process. Make a separate allowance for inflation on a long-term project. No one can say for sure what inflation will actually be. By identifying a separate item, you will have a basis to request a change from the customer if the actual inflation rate differs. The people who end up doing the work may not be the same ones that you anticipated. The person who could have done the work in 80 hours may be gone, only to be replaced by someone who will take 120 hours. Don't be timid about contingencies. If the contingency is inadequate, you'll have to negotiate a contract change every time a problem arises. Project administrative costs are often underestimated. Monthly reports, meetings with the customer, and coordination of subcontractors and team members often seem to take more time than estimated. Project management costs often are 10% to 15% of overall project costs, sometimes even more. Costs to close out a project are also often underestimated or not included in the project budget. Some concerns about closing out a project are discussed in Chapter Seven.

Figure 4-6 appears on page 104   As with schedules, involvement of the project team in preparing the budget is essential. Remember: no involvement means no commitment. A checklist of items to consider when preparing a project budget is presented in Figure 4-6. Use the checklist now to see if there are any gaps in your current project budget.

Exercise

Figure 4-7 appears on page 105   Select a recent project where significant schedule or budget issues occurred. Working with your project team and using the worksheet in Figure 4-7, compare the causes of these issues with those shown in the scoping checklist (Figure 4-1), the scheduling checklist (Figure 4-5), and the budget checklist (Figure 4-6). Identify which items in Figures 4-1, 4-5, and 4-6 were not addressed. Can the issues that arose during the project be related to these discrepancies? Highlight these items in a copy of Figures 4-1, 4-5, and 4-6 and others that seem to be typically overlooked. Refer to these highlighted checklists when you next prepare a budget and schedule for a project or task.

Figure 4-1   *Project Scoping Checklist*

DOs

☐ List all work to be done by your team.

☐ Provide breakdowns of differing types of work in separate sections.

☐ List number and content of deliverables for each task and tie to schedule.

☐ If the project involves collection of information, define whether you will be using available data or developing new information. Define the nature, extent, and sources of available data to be used.

☐ If the project involves exchange of computerized information with the customer or the production of a product involving the customer's use of related software, define what software is to be used.

☐ State the number of alternatives to be evaluated.

☐ If the project involves developing estimates of costs, define the basis and accuracy of cost estimates at different stages of the project (e.g., planning level, design 30% complete, design 50% complete, etc.).

☐ List any fieldwork to be done.

☐ If the project involves the review of draft reports or other intermediate work products by the customer, define the number of times revisions will be made (e.g., one draft, one set of review comments, one final report). If more than one individual within the customer's organization is providing review comments, require that the customer consolidate and reconcile their internal comments.

☐ Define time periods as calendar days or work days after notice to proceed or some other defined point (e.g., complete final report 14 calendar days after receipt of review comments).

☐ For multiphase projects, define the starting and ending conditions of each phase.

☐ Include project administration and management as separate tasks.

☐ Include preparation of a project management plan as a distinct deliverable as part of the scope.

☐ Identify the specifics of the project quality control procedures.

☐ List reporting requirements (e.g., monthly written reports of progress, financial status) and the required content and format of such reports.

☐ If the project involves periodic meetings, define the maximum number of meetings.

☐ State all assumptions in scoping, scheduling, and budgeting the project (e.g., what data are available, what services if requested will be considered scope changes, number of meetings, number of copies of reports to be provided).

☐ Have the written scope reviewed in-house before sending it to the customer.

☐ Define phases if the scope is likely to change based on subsequent work. For example, if certain tasks have to be initially accomplished before the overall scope can be established (e.g., must define existing available data before determining additional data needed and defining the resulting analytical approach), establish a budget and schedule for these initial tasks as Phase One. An output from Phase One will be the definition of the scope, schedule, and budget for subsequent phases.

☐ Include the minimum measurable criteria for acceptance of work.

☐ Define duration of customer review or decision times.

☐ If you are using outside consultants or contractors to help you with the project, define deliverables, schedule, and responsibility for each task to be performed and require regular progress reports to be submitted.

☐ Check to be sure scope is responsive to all stakeholders' needs. Submit a draft to the customer and then meet to discuss the customer's reactions and suggestions.

continued on next page

Figure 4-1   *Project Scoping Checklist (continued)*

DON'Ts
☐ Use ambiguous or open-ended statements (e.g., "including but not limited to," "as required").
☐ Use superlatives and terms of perfection (e.g., "exceptional," "optimum," "all," "best," "highest," or "peculiar").
☐ Use "guarantee," "warranty," "assure," "ensure," "insure," "promise," etc.
☐ Use vague phrases (e.g., "characteristic," "typical," "representative," "approximately," "at least," or "nearly").
☐ Include extraneous requirements (such as collecting data that won't be used).
☐ Include work that is not required to achieve the project objective.

FIGURE 4-2 *Example Form for Scope Change Request and Approval*

Customer:	Change No.:
Location:	Date:
Project:	Requested by:

DESCRIPTION OF CHANGE:

JUSTIFICATION FOR CHANGE:

EFFECT ON SCHEDULE:

ESTIMATED ADDITIONAL COST:	
Work hours	
Labor cost	
Travel	
Computer expense	
Misc. expense	
Contingency	
TOTAL	$

Your Organization		Customer	
Signature	Title	Signature	Title
☐ Approved ☐ Not approved Date:		☐ Approved ☐ Not approved Date:	

FIGURE 4-3 *Project Planning Workshop Checklist of Potential Topics*

☐ Project overview

☐ Project goals

☐ Scope of work

☐ Team members' Myers-Briggs style preferences and how they can affect team performance

☐ Project structure (structure of team/task managers)

☐ Individual responsibilities

☐ Team coordination/communications (including multioffice coordination)

☐ Customer coordination requirements

☐ Availability and location of reference materials

☐ Schedule of milestones and deliverables (includes critical path and internal deadlines)

☐ Project and task budgets

☐ Products or deliverables

☐ Project cost accounting/reporting

☐ Project filing system

☐ Production coordination

☐ Special project challenges

☐ Unique requirements

☐ Schedule for future project meetings

☐ Task managers—first assignments of conducting detailed planning of their tasks, including:

    ☐ Task staffing needs: people, hours, and schedule

    ☐ Task schedule

    ☐ Task technical requirements

    ☐ Products/deliverables

FIGURE 4-4 *Project Work Plan Checklist*

- ☐ Project description, goals, background
- ☐ Project team roster: include addresses, office and cell phone numbers, e-mail addresses, and fax numbers for each team member
- ☐ Scope
- ☐ Task descriptions
- ☐ Task responsibilities: list person responsible for each task
- ☐ Schedule: include milestones and deliverables (see Figure 4-5 for schedule checklist)
- ☐ Budget by task: include cost assumptions (see Figure 4-6 for budget checklist)
- ☐ Sample cost reporting sheets with related explanation of content
- ☐ Invoicing, payables, and status reporting procedures (including the desired invoice format for any subconsultants or subcontractors)
- ☐ Communication protocol between client, team, consultants, and subcontractors
- ☐ Procedures for changes in scope, budget, or schedule
- ☐ Documentation and filing procedures
- ☐ Quality control procedures
- ☐ Health and safety procedures/responsibilities
- ☐ Standards (your organization's and the customer's)
- ☐ Contract provisions
- ☐ List of deliverables: include items such as progress reports, drafts, and prototypes
- ☐ Work that, if requested, will be a change in scope
- ☐ Design criteria and specifications/standards for the end product

Figure 4-5 *Scheduling Checklist*

DOs

☐ Select the type of schedule appropriate to the size and complexity of the project and customer needs. In general, small- and medium-sized projects are best scheduled using a spreadsheet or Gantt chart. Large projects or projects with numerous interdependencies and tight time requirements may require use of a Critical Path Method (CPM)-type schedule. Pick one that can be easily and regularly updated so that it doesn't become wallpaper.

☐ Check with the customer to determine if they have a preferred scheduling method and software package.

☐ Verify the number and experience level of personnel required to meet the schedule based on overall size, project milestones, and interrelated tasks.

☐ Check on the availability of personnel before scheduling individual team members to work on project tasks.

☐ Coordinate project schedule with project team members' individual schedules.

☐ Schedule work realizing personnel will probably not be available to work on your project full-time due to commitments on other projects, business development or other activities, and personal leave requirements. Allow contingency time for shifting priorities and schedules.

☐ If graphics, design plans, or reports are involved in the project, obtain an estimate from these support sections on the amount of time the work will take and incorporate into the project schedule. Allow enough tail-end contingency time for graphics, word processing, printing, review, etc.

☐ Keep the support staff updated on schedule changes that may affect their work.

☐ Provide adequate start-up and review time.

☐ Schedule quality control reviews and identify personnel to perform reviews.

☐ Confirm that the project schedule meets customer's schedule requirements.

☐ Determine the basis for the customer's schedule requirements.

☐ Consult with management when scheduling changes occur; provide updated workload projections.

☐ Review and update the schedule in accordance with contract conditions and as required by project progress. Document reasons for changes to support requests for budget and/or schedule changes.

☐ Coordinate the schedule with subconsultant or other outside contractor/supplier tasks.

☐ Schedule customer and other review times so they are a part of the project schedule.

☐ Notify team members and support staff when changes are made to the schedule.

☐ Schedule a closeout debriefing meeting with the team, as well as a customer debriefing meeting.

☐ Verify that the schedule meets the contract requirements.

DON'Ts

☐ Schedule the project without first coordinating with team members, and managers.

☐ Overschedule individual team members.

☐ Assume your project is the no. 1 priority for staff resources.

☐ Wait to schedule support services. Schedule these services as soon as you know when an item is promised to the customer. This could be months in advance.

☐ Assume everyone understands the schedule. You need to review it in detail with the team, management, and the customer to get concurrence.

Figure 4-6   *Budget Checklist*

Labor for the following (including any appropriate indirect cost or overhead factors):
- ☐ Technical work
- ☐ Fieldwork
- ☐ Status reports
- ☐ Budget reviews
- ☐ Staffing coordination
- ☐ Management
- ☐ Editing
- ☐ Graphics preparation
- ☐ Computer support
- ☐ Health and safety reviews
- ☐ Quality control reviews
- ☐ Technical and financial reviews
- ☐ Beta testing (if the project creates a product)
- ☐ Word processing
- ☐ Proofreading
- ☐ Administrative support/clerical
- ☐ Travel time
- ☐ Meeting time (progress meetings, customer briefings, other presentations)
- ☐ Anticipated salary increases

Costs of the following:
- ☐ Couriers
- ☐ Computer time
- ☐ Graphic supplies and outside services
- ☐ Mailing/postage
- ☐ Shipping
- ☐ Specialized services (e.g., outside laboratory work)
- ☐ Travel, per diem
- ☐ Printing (internal and external)
- ☐ Special items
- ☐ Number of drafts
- ☐ Number of copies
- ☐ Number of pages or sheets
- ☐ Oversized exhibits
- ☐ Equipment lease or rental
- ☐ Equipment purchase (if part of budget, it belongs to the customer)
- ☐ Subcontractor(s) and consultant(s)
- ☐ Other outside services
- ☐ Reference literature/maps
- ☐ Film processing

Other costs:
- ☐ Administrative handling and management of subconsultants, contracts, work products, and risk liability
- ☐ Bond or insurance requirements
- ☐ Contingencies
- ☐ Specialized training required for the project
- ☐ Profit (fee) if applicable

Figure 4-7   *Scope/Budget/Schedule Analysis Worksheet*

PROJECT NAME

ORIGINAL BUDGET

ACTUAL COST TO COMPLETE

ORIGINAL DUE DATE   ACTUAL DATE OF COMPLETION

What scoping/budget/scheduling items listed in Figures 4-1, 4-5, and 4-6 were not originally addressed?

1.

2.

3.

4.

Were there causes of problems other than those listed in Figures 4-1, 4-5, or 4-6?

Yes_____   No_____

(If yes, add these other causes to your checklists)

What can we do differently for the next project to avoid these issues?

1.

2.

3.

FIVE

Who's Pulling What?

Delegating

Why Delegate?

Effective delegation is an essential element of effective project leadership. No matter how strong, fast, and intelligent the lead dog may be, the lead dog alone will never win the race. To succeed, all team members must know their roles and eagerly pull their fair shares of the load. A study of 500 groups of managers by Ohio State University found that those rated as good or excellent leaders were the ones who made the greatest use of delegation (Mackenzie, 1972). Leaders who scored poorly were ineffective delegators. Aggressive, rising stars will be driven away by a manager who refuses to let go of challenging work. When he was chairman of Avis, Robert Townsend emphasized the need to delegate "as many important matters as you can because that creates a climate in which people grow." The energy that people devote to a project increases tremendously when they feel that their skills and knowledge are growing as a result of their project assignments.

Here's a brief quiz to check out your current delegation skills. Check any of the following statements that apply to you:

- ☐ You're working longer hours than your staff.
- ☐ You're taking work home regularly.
- ☐ You're constantly rushing to meet deadlines.
- ☐ You're doing work that you "delegated" to your staff.
- ☐ You're regularly being interrupted by questions about work you've delegated.
- ☐ Your top priority items remain undone.
- ☐ You're the only project manager you can identify for the next big project.

□ You're experiencing low morale or poor initiative on your staff.

□ You're experiencing a high turnover of "rising stars" in your organization or on your teams.

If you checked any of the above items, your delegation skills need improvement. The last point reinforces the folly of not delegating in order to protect your staff from too much work. Eventually you will be left with only those team members who aren't looking to grow and who are more likely to be mediocre performers, and you will end up carrying even more of the load.

As the next part of this quiz, check any of the following benefits that could occur if you were the perfect delegator:

□ Extend the results beyond what you can do alone.

□ Release some of your time to work on new challenges rather than on tasks you can do blindfolded.

□ Develop others, which would lead to greater enthusiasm on their part.

□ Increase involvement of those close to the action where additional facts, expertise, and differing perspectives are available.

□ Speed up decisions.

Now complete the quiz by checking the reasons you are not yet that perfect delegator:

□ I can do it better (or faster).

□ I just didn't think of it.

□ My team is already too busy and I protect them by doing it myself.

□ I'm too busy to delegate.

□ I want the respect of my staff that I get from doing it myself.

□ I'll look weak or incompetent.

□ It makes me uncomfortable.

□ I don't have time to see what could be delegated.

□ I will have to do most of the task before I could explain it to someone else.

An effective project leader would readily check all of the benefits in the second part of the quiz and none of the excuses in the third part. Managers who are not yet leaders agree with the benefits of delegation at an intellectual level, but then they also check several of the reasons for not delegating and they may even add a few to the list. They are not buying into the benefits at an emotional level. Emotional buy-in is essential because many of the problems with delegation come from a deep-seated connection between control and a belief about a person's self-worth. Many people define their self-worth by comparing themselves with others. When they see others gain power, information, or recognition, they experience a sense that something is being taken away from them, diminishing their self-worth. Poor delegators subconsciously fear that a subordinate taking the initiative will make them appear weak and more vulnerable (Covey, 1999). Effective delegators have an inner security that enables them to relinquish control, encourage the growth and development of those around them, and enjoy the success of others.

Managers relate their self-worth to recognition they gain from doing a task, believing "I am what I do." In contrast, leaders reinforce their strong sense of self-worth by delegating work that helps those on their teams grow and develop. Clearly, if you don't break out of the poor-delegator trap, there is a very real limit on what you can accomplish and you will remain a manager, not a leader. You must train your replacement or you'll never be free to move on. Not delegating also sends a signal that you have low expectations of your team's abilities. Performance by others improves as your expectations of them and of yourself rise.

How Delegation Addresses the Triangle of Needs

Effective delegation addresses all sides of the Triangle of Needs:

- ▶ *Be very clear about the task being delegated.* Provide a written definition of the results you expect, when you expect it (including intermediate milestones if appropriate), what the budget is, and who is responsible for the task. Describe the results expected. Set a realistic deadline to which all parties agree. Explain the importance of the task. General George Patton said it well: "Never tell people how to do things. Tell them what to do and they will surprise you with their ingenuity."
- ▶ *Make sure the tools and assistance needed are available.* People need to know from where and how to get tools and assistance. Provide example reports or drawings or identify key people to use for advice or information.
- ▶ *Match the capabilities of the people to the task. Be reasonable in what you expect.* It is stimulating to delegate a task that is a stretch but doable; it is demotivating to delegate a task that is beyond the capability of the delegatee.

- ▶ *Follow up.* Effective follow-up is an essential ingredient in effective delegation. With too little follow-up, the work can wander off track or fall behind schedule with little opportunity to make the needed corrections in an effective or timely way. With too much follow-up, the delegatee may feel smothered and untrusted. It is important to agree on the procedure and schedule for follow-up so that the delegatee does not feel that you are looking over his/her shoulder or that you do not trust him/her. Get the delegatee's input and reach agreement on what and when follow-up should occur. When you do follow up, give latitude to use imagination in accomplishing the work. Be available for questions.
- ▶ *Ask staff to bring you solutions, not problems.* If you allow staff to bring you problems, the phenomenon of "reverse delegation" can result. You'll end up doing the very work you delegated! If staff bring you a problem, ask for their recommendations or ask them to define alternatives or suggest approaches for resolution.

RELATIONSHIP

*Figure 5-1
appears on
page 122;
Figure 5-2
appears on
page 123;
and Figure
5-3 appears
on page 124*

▶ *Delegate the authority.* It is common to want to delegate responsibility and keep the authority. Effective delegation involves just the reverse—delegating the authority but keeping the ultimate responsibility. President Harry Truman's belief in this principle was stated in a sign on his desk that read "The Buck Stops Here." Those doing the task must control the factors determining their success if delegation is to be effective. Recognize that delegating authority is not all-or-nothing. The Responsibility–Authority Matrix (described later in this chapter and illustrated in Figures 5-1, 5-2, and 5-3) shows how to delegate varying levels of authority depending on the complexity of the task and the abilities of the delegatee.

▶ *Have high expectations.* Express your confidence. The delegatee's performance and your expectations are directly related. The effect of expectations is demonstrated by a situation involving 105 Israeli soldiers who were participating in a combat command course (Robbins, 2002). The four instructors of the course were told that one-third of the trainees had high potential, one-third had normal potential, and the potential of the rest was unknown. In reality, the trainees were randomly placed into these categories. The three groups should have performed about equally because they were randomly placed. However, the trainees whose instructors were told they had high potential scored significantly higher on objective achievement tests, exhibited more positive attitudes, and held their leaders in higher regard than the others. The instructors of the supposedly high-potential trainees got better results because the instructors expected it. Leadership behavior that contributes to improved performance includes emotional support through nonverbal clues, more frequent and useful feedback, and more challenging assignments. Leaders expect high performance from their team members, tell them so, and show by their behavior that they believe in their team members—and they get higher performance as a result.

▶ *Reward success through recognition.* Be generous with your praise for a job well done. The praise should come immediately after the good work has been done, not a year later in an annual performance review. The praise should be specific. For example, effective praise would be: "You did an especially good job in completing the outline for the report. I especially liked the fact you included a section on methods to measure customer satisfaction. The fact you completed it a day ahead of time really helped me prepare for my meeting today." Ineffective praise looks like: "Great job. Keep it up." The praise must be genuine. Insincere praise is quickly detected and detested.

▶ *Provide constructive feedback.* Effective delegation involves giving the delegatee an opportunity to grow by involving the delegatee in a task that stretches his/her abilities. An effective leader acts as a mentor by providing the delegatee constructive feedback when the delegatee needs to make adjustments to help advance his/her skills.

Constructive feedback, just as is the case with positive feedback, needs to be made in a timely fashion, not several days or weeks after the work in question has been done. Constructive feedback should first recognize the positive aspects of the delegatee's work, point out the area for potential improvement, and make it clear that this is your perception. Explain your rationale and work together to come up with a way to achieve the needed change. For example: "You did an especially good job in completing the outline for the report. I especially liked the fact you included a section on methods to measure customer satisfaction. I think it would increase the value of the report to our readers if the section were expanded to discuss the advantages, disadvantages, and relative costs of these methods. I think that information would allow the readers to make a better judgment of how the methods might work for them. Do you agree? How would you approach gathering and analyzing the information to add this material?"

Six Steps to Effective Delegation

Step 1: Choose a capable person.

Evaluate both the assignment and the person. Have reasonable expectations about what the person can do. Be sure you are not delegating too-much–too–soon, or to someone who may not be able to complete the task properly. Andrew Carnegie said it well: "The secret of success is not in doing your own work, but in recognizing the right person to do it."

Step 2: Explain the objectives.

Make sure the person clearly understands and accepts what is being delegated. Clearly define desired results and lines of authority, allowing enough flexibility for the person to use his/her initiative. Show how the task fits into achieving the overall project objectives and how it relates to tasks being done by others. Exceptional levels of individual motivation were observed in the multitude of contractors involved in the mission to put a man on the moon by the end of the 1960s. For example, the individuals charged with developing the new material needed for space suits understood that their efforts were going to put a man on the moon and put the United States back in front of the Russians in the "space race" for scientific superiority. Imagine the difference in motivation if these individuals had been told only the technical specifications that the material must meet without understanding the role the material would play in such an important mission. Yet, all too often we see instances where project team members are given only their task assignments without being given the information on how their tasks fit into achieving the overall project objectives.

Step 3: Give the person the tools and authority to do the task.

Identify the level of authority to be given to the delegatee (as in the Responsibility–Authority Matrix, described below). Clearly define the boundaries and authority of the assignment. This should include schedule, budget allocated, resources for assistance, and the single and overlapping areas of authority. Those affected by the delegatee's authority need to be notified.

Step 4: Follow up, keep in contact.

The delegator is still responsible for the delegated task. Set up methods for monitoring the delegatee's work, including periodic reporting and review sessions, ahead of time so the delegatee feels supported but not suffocated. Ask yourself, "Am I really letting the other person do it, or am I keeping strings attached and withholding authority, thus hampering the person's freedom to take proper action?"

Step 5: Accept other approaches.

The delegatee may not perform a task the way you would, but if his/her approach achieves the objectives, accept the new approach. If you try to change it to your way, it will defeat the purpose of delegation, and the person may be discouraged from taking the initiative in the future. Have you ever had a supervisor review something you have written and cover the pages in red ink, not because your supervisor has a problem with content but prefers a different style of writing? You may be less motivated to make your best effort on the next writing assignment knowing that, regardless of the excellence of content, your supervisor is going to cover the pages with a multitude of subjective comments. This may lead to reinforcement of a supervisor's unwillingness to delegate because it wasn't done well. This is a vicious cycle.

Step 6: Acknowledge and recognize.

Give the person positive feedback for successes and constructive feedback for course corrections to help them grow in skills and confidence.

Personal style preferences affect how easily and effectively each of the above steps is carried out. The relationship between these preferences and delegation is discussed later in this chapter.

The Responsibility–Authority Matrix

A key to effective delegation is to recognize that authority can be delegated in differing degrees. It is essential that the delegated responsibility and delegated authority match the delegatee's abilities. When a delegatee's responsibility exceeds his/her authority or ability to act, the result will be unrealistic expectations, wasted time, and a lot of frustration for all concerned. Consider the following four broad levels of increasing delegated authority:

Level 1: Look into the problem	Report all of the facts to me and then I'll decide what to do.
Level 2: Develop a recommendation	Let me know alternative actions, including pros and cons of each, and recommend one for my approval.
Level 3: Develop an action plan	Let me know what you intend to do; don't take action until I give final approval.
Level 4: Take action	Gather and analyze the information, take action, and let me know what you did.

Figure 5-1 appears on page 122

Figure 5-1 presents a Responsibility–Authority Matrix example that will help guide you in applying the concept of delegating appropriate degrees of authority. In this example, Alexa Brown desires to delegate the writing of the next project report to Jamie Smith. Alexa informs Jamie that he will be preparing the next report. Jamie has been working on the project, has prepared some interim progress reports, and is eager to try writing the next official project report. Alexa knows that Jamie has done some shorter, simpler reports; she is not ready to give Jamie Level 4 authority but is comfortable with delegating Level 3 authority. Alexa expresses confidence to Jamie that he will do an excellent job and that she would like to take a look at Jamie's draft before it is finalized. After Jamie agrees with the level of authority being delegated, Alexa suggests that they discuss the details of the delegated work and agree on what would be appropriate follow-up. They agree that preparation of an outline is the first step and that it would be logical for Alexa to follow up by reviewing the outline. They agree that July 15 is a reasonable time to complete the draft and that both expect the outline preparation will take about four hours of effort by Jamie. This information is recorded as Checkpoint 1 on the example matrix (Figure 5-1).

They continue the discussion, identifying the other key checkpoints shown on the example matrix, agreeing that Alexa will provide a review at each checkpoint. This is not a "to-do" list but rather a list of the checkpoints in the task delegation when reviews, potential course corrections, and status can be discussed. Each keeps a copy of the completed matrix. By discussing the task and completing the matrix, both the delegator and delegatee can be confident that their perceptions of the delegated task are in agreement. There is a clear alignment of responsibility and authority as well as agreement on the appropriate follow-up. Alexa can use the copy of the matrix as a reminder of when to check in with Jamie.

The matrix also serves as a useful guide to structure the discussion of the task as it focuses both parties on identifying the authority being delegated, ensuring that the level of authority and responsibility are in agreement, ensuring that the delegatee's abilities and responsibilities are in agreement, and that the essential elements of the delegated task (including level of effort and schedule) are agreed upon.

Project Learning Plans

A project leader knows that team members work with much greater energy when their project assignments align with a personal growth goal. At the time the project is being planned, ask each team member if there is a task on the project that would advance one of his/her personal goals. For example, the project may involve the application of software that one of the team members is eager to learn. Delegating a task to the team member that involves applying this software will unleash personal energy that would have remained bottled up if the team member had been delegated a task he/she has done several times before or that does not align with a personal growth goal. There is a temptation to give the team members the same tasks they have done before because they have directly relevant experience and perhaps can do the task more efficiently than anyone else. However, an "anyone else" out there may be eager to learn the task and would do it with a great deal of enthusiasm and energy. The short-term inefficiencies of delegating the task as a learning opportunity may be offset by the fact that you now have two happy, energized team members, each doing tasks that are helping them grow—and you are creating two team members with more capabilities. It is not always practical to meet a team member's personal growth goals, but be aggressive in looking for opportunities.

Personal Styles and Delegation Challenges

Each personal style preference brings its own unique set of challenges and benefits to being an effective delegator or delegatee. Some delegation challenges and solutions are discussed next.

	Challenges as delegator
Extraverts	▶ Need to recognize that they must stop talking through their ideas about the task with others and delegate it to a specific individual or individuals in a timely way so that the work can be completed on schedule.
	▶ Need to take special care in describing the scope of the delegated task because of their preference to think out loud and bounce ideas off of others as part of their thought processes. Need to recognize that this brainstorming may be interpreted by the delegatee as part of the assignment when, in fact, the Extravert may have been just throwing out ideas without really wanting to carry them forward. At the conclusion of the discussion of the delegated assignment, the Extravert and delegatee need to summarize the scope as they each see it, checking the perceptions of the delegatee to be sure that inappropriate parts of the brainstorming are not erroneously translated into part of the assignment.
	▶ Are likely to look forward to the personal contacts and discussions with delegatees involved in following up with them. However, when following up, Extraverts need to be careful not to throw out other ideas just for the sake of talking about them—the delegatee may think that they have just been given new direction.
	▶ Are typically more comfortable than Introverts in praising others and in recognizing and acknowledging a good effort, but must take care not to be so involved in discussing their own ideas that they miss opportunities to give praise.
Introverts	▶ Sometimes find the delegation process uncomfortable because, to be effective, it involves initial and follow-up face-to-face meetings with others. Need to avoid the tendency to think about the task to be delegated, then send a written task description to the delegatee, and finally rely on written progress reports as follow-up. Although the face-to-face interactions tap into their energy reserves, Introverts need to recognize that discussing their thoughts about the task and checking on progress by interacting with the delegatee will ultimately result in net gains.
	▶ Have a tendency to share only their conclusions about a task to be done without sharing how they reached those conclusions. This can leave the delegatee wondering about the underlying reasons for the task and how it can best be approached. Introverts need to share their thought processes.
	▶ May resist delegating a task because they think that they will have to do most of the task before they can explain it. Introverts often think through and do a task before they feel comfortable talking to others about the task. Need to overcome this tendency by recognizing that it is acceptable to talk about the task before thinking it through.
	▶ Need to recognize that it typically will take a conscious effort for an Introvert to praise others. Need to see that earned praise is given.

	Challenges as delegator
Intuitives	► Are often very effective at describing how a task fits into the big picture of the project and at getting the delegatee excited about how the task can advance the project. This may give the delegatee only the broadest of pictures of the specific task to be done. While effective delegators do not spell out every detail and leave latitude on how the task is to be done, they do give enough detail about the task that it is reasonable to expect the desired result will occur. Intuitives need to make a conscious effort to provide the right amount of detail. ► May feel that they don't have the time to examine the project or task to see what parts could be delegated. The true cause for this discomfort may be the tendency to look at the big picture and not the detailed parts, no matter how much time they take. Intuitives need to combat this tendency by working with the delegatee to look at the parts of the task that can effectively be done by the delegatee. ► Are eager to try new approaches and test out new theories, but need to balance this tendency with a clear recognition of budget and schedule constraints and the need for a workable result as they work through the approach with the delegatee.
Sensors	► Are usually very effective at working with the delegatee to structure a logical, step-by-step approach to the delegated task. However, they may have a tendency to give the delegatee too much information, which does not give the delegatee enough latitude to grow as a result of the assignment. Sensors need to limit the amount of detail they provide to that needed to define the follow-up procedures, outputs, budget, and schedule for the task without providing every detail that is on their minds. ► May fail to show the delegatee how the task fits into the big picture of the project. Sensors need to show how the work to be done by the delegatee will advance the overall project. ► May feel more comfortable relying on the way a similar task has been done in the past. Sensors need to make the delegatee aware of past work that may be useful as a resource, but avoid prescribing past methods as a way to do the task. ► Must resist their tendency to micromanage, because it takes away the delegatee's initiative.

	Challenges as delegator
Thinkers	▶ May struggle with delegation because they believe they can do the task better than other team members. Thinkers need to realize that they will not free up time for themselves unless they assist others to develop, and that others may have the ability to do the task as well (or, heaven forbid, even better) if given the opportunity.
	▶ May struggle with a tendency to not want to appear weak or incompetent by asking someone else to perform the task. Thinkers need to realize that team members will view them more positively rather than more negatively if they delegate effectively.
	▶ Even though they may be interested only in a stimulating (to them) intellectual probing of why the delegatee is doing what he/she is doing, Thinkers need to realize that this may appear to others to be very critical of the work. This tendency will interfere with effective delegation. Thinkers need to temper this tendency and offer constructive feedback to the delegatee when appropriate.
Feelers	▶ Tendency to seek harmony and avoid conflict may come back to haunt them if they pay more attention to the human aspects than the details of the task.
	▶ Must balance their need to solicit and consider the desires of team members—about which project tasks may be in alignment with their personal growth goals with the realities of the task budget and schedule—even if it means disappointing some team members on the immediate project.
	▶ May believe that, by doing tasks themselves rather than delegating, they are protecting their team members. This makes them prone to reverse delegation, in which they take on a delegatee's problem. This tendency may actually drive away the best team members, who sense they are being stifled in such an environment. Rather than feeling protected, the team may look upon the delegators as keeping all of the interesting work for themselves.
	▶ In their desire for harmony, Feelers may avoid offering constructive feedback, fearing potential confrontation. They need to overcome this concern by looking at such a situation as a chance for stimulating discussion of alternate ways to improve the task.

	Challenges as delegator
Perceivers	► Often are "too busy" to delegate or claim they just didn't think of it. They may lack a sense of urgency and may procrastinate, mulling over a task and not delegating it until it has become a crisis. Perceivers need to address these tendencies by being aware that it will take a conscious effort to do the necessary advance planning, and by accepting that such planning has great value to them and their team members.
	► Must recognize that they need to step away from the tendency to go with the flow. Instead, they need to focus their energy on developing a structured approach—including appropriate delegation—to planning the project.
	► Like to consider all options, a useful preference in discussing how to approach a task as long as they recognize that the task will never be completed unless the options are limited to the most relevant ones. When they discuss the task to be delegated, Perceivers need to be aware of their tendency to want to discuss options and of how this tendency can get in the way of reaching closure. Because of the large number of options Perceivers may generate, the delegatee may become confused.
	► Need to set up an effective reminder system to follow up with the delegatee on the agreed-upon schedule.
Judgers	► As a result of their desire to get to closure, may have a tendency to give orders rather than truly delegate, resulting in a tendency to be too directive.
	► May look upon the time required for effective delegation as a delay in completing the task and, as a result, believe that it would be quicker to do it themselves. Judgers need to address this tendency by taking the time to explore options with the delegatee before finalizing the delegation, by asking the delegatee if he/she has any other options that should be considered, and by giving the delegatee latitude on how the work will be carried out.
	► Must recognize that others may complete the task in a timely way but using a less structured approach than a Judger may have used. In such circumstances, Judgers need to be patient with the delegatee's less structured approach. They also must resist the temptation to impose their own approaches in an attempt to speed up task completion.
	► Need to be sensitive to the agreed-upon procedure and schedule for follow-up and resist the temptation to follow up even more frequently just to be sure the delegatee is sticking to the plan.

Circle your style and write down some of the delegation action steps from the preceding table that you could take to improve your delegation skills.

E–I

_____

S–N

_____

T–F

_____

J–P

_____

	Challenges as delegatee
Extraverts	▶ Must exert the effort to carefully listen to the delegator. The desire to express their own ideas about the task may interfere with understanding what the delegator has in mind. ▶ Avoid the tendency to talk through their own ideas; rather, they should check their perceptions of what they have heard by asking questions, such as, "If I'm hearing you correctly, you want me to. . . . Is that correct?" ▶ Cannot assume the delegator, who may be an Introvert, is sharing all of his/her thoughts about the task. To fully understand the assignment, need to ask questions to draw out the delegator's thoughts. ▶ Need to clarify their understanding of which aspects of the assignment are essential. This will avoid their treating all ideas that may arise during the discussion as equally important when they might be just interesting points to discuss.
Introverts	▶ Need to ask questions they have about the assignment at the time they are interacting with the delegator, although they may want to think about the task before asking. Asking can avoid the waste of energy and time spent internally processing information that potentially is irrelevant. ▶ Need to seek out the delegator if questions arise or circumstances change so that the effects on the task can be discussed in a timely way. Relying on e-mail or a memo for communication about changes can lead to misunderstandings. ▶ Need to share their thoughts on how they are going to approach the task, rather than think the task through to a conclusion before discussing their methodology. ▶ Need to ask questions to be sure that there are no unsaid comments and concerns that need to be explored. Unless asked, the delegator, especially if also an Introvert, may consider them too unimportant to mention.
Intuitives	▶ Need to understand what details are essential to successful completion of the task. Intuitives need to ask enough questions to get the detailed information needed to avoid solving a problem that interests them but that is the wrong problem. ▶ Need to listen carefully to and understand the delegator's thoughts about the steps involved in a successful approach to the task, even though they may think the detailed steps will be obvious and need not be discussed. ▶ Should check their ideas on alternative approaches or task changes with the delegator before spending time pursuing them. ▶ Need to be sure that information from the past that would be helpful is fully considered.
Sensors	▶ Need to ask questions to establish the relative value of various objectives that an Intuitive delegator may discuss and to clarify ambiguous terms that an Intuitive may use. Need to check perceptions of priorities. ▶ Even though eager to get to work on the task, need to understand how the project fits into the delegator's future needs and the desired result before beginning work. ▶ If the delegator is also a Sensor, need to be careful to see that more than short-term solutions have been considered.

Challenges as delegatee	
Thinkers	▶ Should not hesitate to ask questions to clarify the assignment. Need to set aside concerns about appearing weak or incompetent by asking questions.
	▶ Need to be sure that the discussion with the delegator includes how input is to be obtained from those that may be affected by the task.
	▶ Need to guard against altering the delegator's needs to more closely match their perceptions of the right approach. Also must guard against applying their favorite solutions to the delegator's needs, whether appropriate or not.
	▶ Need to remember that, when discussing the task, not everyone enjoys pursuing a point just for the sake of argument.
Feelers	▶ Need to remember that it is acceptable to have differing views about how to approach the delegated task and that discussion about such differences in opinion can lead to a better approach.
	▶ Need to remember that disagreements about approach are not directed at them personally.
	▶ Should realize that a good task approach may be based on some logical steps that may not all be "feel-good" steps.
Perceivers	▶ Need to remember that the delegator has specific goals in mind for the task and that these goals are more than a general guideline for ongoing discussion.
	▶ Need to understand the procedure for addressing any changes in the task scope that they want to suggest.
	▶ Need to understand the task deadline and intermediate milestones and accept the importance of meeting these deadlines in a timely way.
	▶ Should use a "tickler" system to remind themselves of upcoming deadlines and milestones.
Judgers	▶ Should take the time to have a thorough initial discussion with the delegator, and avoid the temptation to rush to start work on the task before fully understanding what is needed for the task to be successful.
	▶ Need to accept that they will have to adjust the approach if the delegator requests changes in the task approach and goals as more information becomes available.
	▶ Should strive to be comfortable with the fact that there is no need to finalize the approach to each step of the work in advance—the approach to some steps may depend upon the results of other steps.
	▶ As the work proceeds, need to call new information to the attention of the delegator, even if it might mean delaying closure of the work.

Have your primary delegatee complete the table below, or complete it yourself if you are in the position of a delegatee. Circle your style preference and then write down the actions that you could take to ensure you receive appropriate and adequate information and direction from the delegator.

E–I

S–N

T–F

J–P

Exercises

Figure 5-2 appears on page 123; Figure 5-4 appears on page 125; Figure 5-3 appears on page 124

1. Identify a task that you would like to delegate and choose an appropriate person as the delegatee. Use the Responsibility–Authority Matrix in Figure 5-2 to structure a discussion with the delegatee about the task. Then, analyze the effectiveness of your approach using Figure 5-4.

2. If you are in the position of a delegatee with a manager who does not delegate effectively, complete Figure 5-3 for a task you want to perform or learn. Write down the checkpoints you would like to use, and estimate the time and effort to reach each checkpoint. Then approach your manager with it. You will end up training your manager!

3. Think of a situation in which you found a delegated task really excited and energized you. What were the characteristics of the task? What was effective about the way that the task was delegated that you could use in the future?

4. Think of a situation in which you felt that you were dumped on rather than delegated to. What were the characteristics of that situation? How can you avoid creating the same situation?

5. How do your personal style preferences contribute to the delegation challenges that you face? What could you change about your behavior to address these challenges?

Figure 5-1   *Example Responsibility–Authority Matrix for Effective Delegation*

YOU AS DELEGATOR:

Your name: Alexa Brown

PROJECT/TASK TO BE DELEGATED

Delegatee name: Jamie Smith

Project/task description	Degree of authority				Level of effort to reach checkpoint (hrs)	Deadline
	1	2	3	4		
WRITE NEXT PROJECT REPORT			X			9/30
Checkpoint 1 Prepare outline					4	7/15
Checkpoint 2 Prepare graphics plan					4	7/25
Checkpoint 3 Prepare each section					20	ongoing
Checkpoint 4 Prepare Executive Summary					8	9/15
Checkpoint 5 Discuss feedback on final report, revise report					4	9/25

Figure 5-2 *Responsibility–Authority Matrix for Effective Delegation Worksheet for Delegator*

YOU AS DELEGATOR:

Your name: _____

PROJECT/TASK TO BE DELEGATED

Delegatee name: _____

Project/task description	Degree of authority				Level of effort to reach checkpoint (hrs)	Deadline
	1	2	3	4		
Checkpoint 1						
Checkpoint 2						
Checkpoint 3						
Checkpoint 4						
Checkpoint 5						

Figure 5-3   *Responsibility–Authority Matrix for Effective Delegation Worksheet for Delegatee*

YOU AS DELEGATEE:

Your name: _____

PROJECT/TASK TO BE DELEGATED

Delegatee name: _____

Project/task description	Degree of authority				Level of effort to reach checkpoint (hrs)	Deadline
	1	2	3	4		
Checkpoint 1						
Checkpoint 2						
Checkpoint 3						
Checkpoint 4						
Checkpoint 5						

Figure 5-4 *Delegation Effectiveness Worksheet*

1. What issues or challenges did you and the delegatee have in agreeing on the following:

    a. Appropriate level of delegated authority?

    b. Appropriate checkpoints and related reviews?

    c. Level of effort involved in the task?

2. How did differences in personal style preferences affect the delegation process?

3. What could you do differently to improve the effectiveness of the delegation process?

SIX

Where Are We?
Are We There Yet?

Monitoring and Adjusting Project Progress

What We Need to Know

If our team wanders from the course, we will surely fall behind and, in a long race like the Iditarod in Alaska, we may run out of provisions as well. The lead dog is constantly scanning what lies ahead and adjusting to stay on the course. The story of Nanook in Chapter One shows how the lead dog will keep the team on track, even when visibility is so poor that the rest of the team has trouble seeing the path to their destination. Imagine how far off course we could be if the lead dog decided to chase every distraction seen along the way. Imagine how little progress we would make if the lead dog ignored the others if each of the team members decided to pull in a direction of its own choosing. The lead dog stays focused on the finish line and quickly pulls the team back into line if they try to veer off in a different direction.

The project leader has to be constantly scanning the horizon to gather information on how the team is progressing and to see what adjustments need to be made to stay on course. The leader measures progress toward the project objectives, evaluates what needs to be done to reach the objectives, and takes appropriate actions. The process is a continuous one to detect and resolve potential problems so that the project stays on schedule, within budget, and produces the end result desired by the customer.

The information the project leader needs to gather goes far beyond the obvious of costs versus budget and schedule compliance. Look at the following list of needed information and see if the information comes from the accounting system or from interacting with others. Also, consider which side or sides of the Triangle of Needs the information falls upon. Add other items of information relevant to your projects.

Information needed to determine project status	Comes from accounting system	Comes from interacting with others	Side(s) of the triangle: Content? Procedural? Relationship?
1. Effort and cost to complete the project			
2. Actual status (percent complete) of completion of project deliverables			
3. Team members' perceptions of the scope			
4. Customer expectations and perceptions of the scope			
5. How does the customer feel about our work to date?			
6. Stakeholder/public perceptions and expectations			
7. How do team members feel about our work to date and their roles?			
8. Problems, perceived obstacles			
9. Availability of resources			
10. Changing priorities			
11. Workload of team members			
12. Schedule and milestones met or missed			
13. Status of quality control reviews			
14. Cost to date			
15. Budget remaining			
16. Are invoices to the customer up-to-date (if it is a revenue producing project)?			
17. Is the customer paying the invoices?			
18. Are you getting the work produced for the charges you are being billed by contractors or consultants?			
Other			

The information to assess items 1 through 13 comes from interacting with the team members or customer, not from the accounting system. Yet one of the most common excuses we hear from project managers (not leaders) is that they can't do a better job of monitoring their projects because their organization's financial accounting system is slow or inaccurate or has some other perceived shortcomings. You cannot get the majority of the information you need to effectively monitor your project progress from your computer screen or a print-out from accounting.

People do projects. Monitoring and controlling projects requires you to communicate with people, measure their progress, and get their commitment to course corrections. Consider the Manhattan Project that produced the atomic bomb during World War II. The project involved thousands of interrelated tasks at locations in different parts of the United States. Yet the project was successfully accomplished within a very demanding schedule without the aid of today's computers and related planning and accounting software. The strong leadership of the project leader, General Groves, coupled with the interaction among his key project team members was the key to the success of the project. A multitude of systems are available today to track costs, staff hours, task interrelationships, and so on. Systems make it easier to organize information about the project but, as in the case of the Manhattan Project, people plan, control, and implement the project.

Ongoing Communication

You must integrate communication skills, experience with other teams, your knowledge of personal style preferences, cost and schedule information, and technical expertise to assess the status of a project, make appropriate adjustments, and keep the sides of the triangle in balance.

A project leader seeks and obtains accurate input, support, and advice of the project team. Personal contact with the team members is essential. Individual meetings with team members, with time spent looking at their work products, provides the team leader with the best information. Such meetings build rapport while also giving you the opportunity to personally see and discuss the status of the work. It is critical that you personally discuss the work under way and form your independent opinion. By focusing on the people doing the work, you gain the knowledge needed to monitor and control the project. Also, your personal interest in their work contributes to a motivating environment. Listed below are keys to effectively recognizing and dealing with personal style preferences for taking in information (Intuitive or Sensing) and making decisions (Feeling or Thinking) so that you effectively communicate.

Communicating with various styles

Communicating with Intuitives	▶ Listen to their ideas and stress your points at the idea level. ▶ Don't expect precise detail. ▶ Don't overwhelm them with facts and figures. ▶ Don't be surprised if you are not hearing complete sentences. ▶ Be prepared for the subject to change without warning. ▶ Discuss opportunities for the future. ▶ Ask: "What if?" "Let's imagine." "Are there other possibilities?" ▶ With Introverted Intuitives, ask questions to get them to verbalize their inner visions. Show you understand their visions. ▶ Be enthusiastic with an Extraverted Intuitive and subtle with an Introverted Intuitive. ▶ Reinforce the need for specific actions that need to occur in the immediate future. Intuitives tend to procrastinate in dealing with the here and now. ▶ Don't expect to hear how long it will take to complete the task or how the ideas will be specifically implemented. Intuitives may get irritated when pushed for details.
Communicating with Sensors	▶ Make a list of topics to cover and stick to it. ▶ Use precise language. Finish your sentences. Don't change subjects in midstream. ▶ Have the facts in hand to support any points you are trying to make. ▶ Avoid making blue-sky suggestions or anything that can't be supported by the five senses. ▶ Remember that brainstorming and visioning is difficult for Sensors. They prefer to engage in some immediate activity. ▶ When put on the defensive, Sensors may gather large amounts of data to support their positions. Avoid getting into a facts battle. ▶ Focus on the trees, not the forest, if you expect the Sensor to relate to you. ▶ Have a clear picture of how to implement your ideas, including a step-by-step plan and possible schedule and cost to complete the task.
Communicating with Feelers	▶ Spend some time talking at a personal level; appreciate some aspect of their work, taking notice of the person as an individual. ▶ Don't start the discussion with criticism or negativity. Comment on what you agree on and what you like about the person's approach or work. ▶ Remember that the lack of harmony will probably adversely affect a Feeler's performance. To a Feeler, a good decision takes others' feelings into account. ▶ Expect the personal aspects of the team and work environment to be significant factors. ▶ Ideas and facts will not be enough. Feelers want to evaluate and relate to the issues.

Communicating with various styles	
Communicating with Thinkers	▶ Don't expect social niceties. Get to the point. ▶ Anticipate criticism or a list of problems. Don't take it personally. ▶ Be factual. Don't get personal. ▶ Use logical analysis and be objective. ▶ Be prepared. Thinkers like well-thought-out ideas. ▶ Remember that arguing one (or both!) sides of an issue may be the Thinker's way of understanding an issue. The arguments may not be their true feelings on the issue. ▶ Show that you hear the person's rational interpretation of the issues and the person's suggested solutions.

Figure 6-1 appears on page 145

Figure 6-1 presents a checklist for effective project communication.

Project Elements that Can Be Controlled

There are six basic elements of a project that you can control:

1. *Scope.* If the scope is set at the start of the project, how can it be a controllable element? Even if it is precisely defined at the start of the project, you cannot assume that everyone working on the project will be working toward the same scope throughout the project. You can exert control so that the team members are all working on the agreed-upon scope. "Scope creep" can be avoided only by effective and frequent communication of the scope to the project team. Your continuous assessment of how the team's efforts relate to the scope is essential. No project goes according to plan, so scope change will likely occur. These changes must be clearly understood by the team so that they are all working toward a current version of the scope.

2. *Time.* You can control the time at which the work is done relative to the overall project schedule and relative to other tasks. The number of people you assign to do a task may also determine the amount of time required to do the task.

3. *People.* You can control the number and type of people working on the project—and when they do their work. If it is going to take 240 staff hours to prepare a report, you can't expect to do it in 4 hours by putting 60 people on the task.

4. *Costs.* Don't confuse tracking costs with project control. You must go beyond tracking costs to analyze the costs in relation to overall progress and take appropriate action to exert control. You may be tracking costs, finding that the rate of expenditure is right on plan, and then discover, when the funds are exhausted (right on schedule!), that the

work is only 75% complete. This sorry result can be avoided by keeping in touch with the people doing the project work to assess their progress in relation to cost.

5. *Quality.* There is no point finishing on time and on budget if the result is a poor quality project. Standards that define quality for the specific project should be established at the start of the project. How progress toward and compliance with these standards will be measured and reported also must be established.

6. *Communication.* Critical communication occurs in at least five ways: (1) from the project leader to the team, (2) from the team to the project leader, (3) among team members, (4) between the project leader (and sometimes from the team members) and the customer, and (5) between the project leader and the organization's upper management. Each of these links is important to project success and must be managed and monitored as carefully as schedule and costs. If the leader doesn't know what the team is doing, even a sophisticated project control system will fail.

Project Reviews

Properly planned and scheduled project reviews allow you to catch deviations from plan and make adjustments before problems arise. Reviews may include ongoing regular status reviews and periodic, structured quality control reviews.

You cannot rely on project team members to alert you to problems because they are reluctant to be the bearers of bad news. People in the midst of doing project work often tend to be overconfident about the time and cost to complete the work. You must periodically and personally review project progress.

We had firsthand experience with a situation that exemplifies the problems that can occur without periodic reviews. We were asked to conduct a review of a project submittal of design drawings and specifications that were reportedly 90% complete. The project manager reported to us that the project budget was in good shape because only $850,000 of the $1,000,000 design fee (85%) had been expended to get to the 90% completion point. We asked all of the technical disciplines to estimate their costs to complete the design. Their estimates of the costs to complete totaled $300,000. The project manager had not conducted periodic reviews with the various disciplines and, as a result, faced a very unpleasant surprise of a potential budget overrun of $150,000. At this late stage of the project, there was little the project manager could do to adjust the work plan to reduce the total design cost. Had the project manager been asking the right questions of the team members on an ongoing basis, the project manager would have been aware of the situation in time to make adjustments.

Reviews can run the gamut from informal exchanges with project personnel to formal reviews at scheduled points in the project. Introverts need to be reminded that there is much to be learned from the informal exchanges they can have by getting out of their offices and talking with the project team. Extraverts need to remember that they will only learn about project status by *listening* to the team members.

Include internal formal review efforts in the project plan as well as any external reviews with the customer. Treat the reviews as small projects within the main project. On projects involving diverse technical disciplines, you may need to involve others in the reviews. Each person involved must know how much time and what detail of review he/she is expected to provide.

A key to reviews is that they be carried out in a friendly manner so that they are not threatening to the project team. If you hear some bad news, don't lapse into blame and recrimination. Rather, ask what in the system allowed the problem to occur. Don't look for an individual to blame. If you attempt to gather information in an authoritarian or accusatory manner, information will be hard to get. By developing rapport, you'll have a much better chance of obtaining productive results. Thinkers need to be especially aware that their directness and tendency to ask very direct questions can be perceived as threatening by some.

When the purpose and need for reviews is properly explained, project team members will welcome the chance for some outside input to their work and the chance to communicate their concerns. Explain to the project team at the start of the project that reviews are an integral part of achieving a quality result. Identify the proposed schedule and budget for any formal reviews that are scheduled, and get the team's agreement that the scheduled times are appropriate. The reviews offer team members an opportunity to learn from the reviewers.

Figure 6-2 appears on page 146

The best way to get information is to ask questions and *listen*. A list of useful questions is provided in Figure 6-2.

If there is some tangible output (a task report, a draft project report, computer outputs, prototype construction, etc.), you should review it. Your own independent assessment of its status is important.

The frequency of contact with your team members must be based on good judgment. It is possible to over-control as well as under-control a project. There have been cases where a manager's contact with the team created a problem where none existed. In one case, a project manager with a preference for Extraverted Intuition loved to walk the halls and fire questions to the staff. The staff complained to us that they kept getting conflicting direction from the manager about the project. The manager complained to us that his staff wouldn't stay focused on the scope and kept charging off on wild goose chases. In the manager's mind, he was simply asking questions to satisfy his curiosity about various possibilities related to the project. The staff was treating the manager's questions as instructions to redirect their efforts toward these other possibilities. After understanding his tendencies and how his every word was given great weight by the staff, the manager was careful to clarify when he was just collecting information and when he was giving direction to team members.

Figure 6-3 appears on page 147

Figure 6-3 presents a project review checklist that can help you structure discussions with team members. It is designed to solicit information on all eighteen of the project status items listed at the start of this chapter. The checklist assumes that the project is funded by an external customer and that the customer is invoiced on a periodic basis for work done on the project. The checklist questions can readily be adjusted for other types of projects. Do not view the checklist as a form to fill out! It is intended to help structure and focus a discussion. Typically, the critical items can be reviewed in a ten- to twenty-minute discussion—one that may save many hours of time later if you need to try to get the project back on track because you did not monitor project status in a timely way.

Project leaders can use the checklist to structure discussions with project managers and project team members on a regular basis. We often pair up attendees in our training sessions and have one of them talk through the status of one of the ongoing projects or tasks using the Figure 6-3 checklist with the other, and then we have them reverse roles. After their discussion, we ask how many realize something about their own project or task that they did not realize before the discussion. Usually, all of the participants raise their hands. Frequently, the participants later tell us that the information gained in the exercise was so valuable that they have set up a "buddy" system with a peer in the organization to periodically have a scheduled discussion about each other's projects using the checklist as a guide.

Once you've gathered the information and made your evaluation of the project status, discuss your conclusions with the individuals on the project team before you make any modifications. If there are problems, they are likely to have some good ideas on what may have gone wrong and how to correct the problems. The more the team is involved in uncovering and solving any problems, the greater chance that the solutions will be accepted and effective.

Another way to use the Figure 6-3 checklist is in regular project meetings where everyone describes the status of his/her tasks. Everyone hears the same information and is aware of how any changes by others may affect their own tasks. The team can identify, discuss, and address potential issues to minimize how they affect the project. If you are working with a team with individuals in different geographic locales, you can use the checklist on a conference call to determine project status. If you are working with several project managers on different projects, the checklist can be used to get a snapshot of the status of each project.

Dealing with Project Changes and Problems

When you find the team is getting off track, it's time to stop and do some analysis. It is important that you establish a safe environment for people to make changes by focusing on what in the system allowed the problem to occur, as opposed to laying blame or seeking the "guilty."

Personal style preferences affect how changes and problems are handled; each preference brings some strengths and some challenges. Listed below are some ways that you can draw upon a range of preferences to be effective in dealing with changes or problems.

How to use your preferences when confronting changes or problems
Extraverts
▶ Engage others who do not seem to be participating in the discussion.
▶ Be prepared to stop talking, even in the middle of making a point, to avoid being redundant when discussing the problem or change.
▶ Control your urge to speak. When you do, restate what others have said and, if needed, add your own ideas but only one at a time.
▶ Do not interrupt others or take issue with their points of view before they have finished speaking.
Introverts
▶ Don't filter your ideas; be willing to share your ideas spontaneously.
▶ Don't rule out any of your ideas or observations of what seem to be problems because they seem trivial to you. Your observations may spark a problem-solving idea by someone.
▶ Don't hesitate to push Extraverts for clarity to find which ideas are important to them and what part of the conversation is just their thinking-out-loud process.
Sensors
▶ Assist others in defining the problem in real and specific terms. Urge others to keep their ideas simple and to the point.
▶ Ask others to use facts. If someone attributes statements about the problem to other individuals, ask for the specifics of from whom and from where they got their information.
▶ Make a conscious effort to piece together facts you are hearing into a bigger picture that will assist in identifying the root causes of a problem or a change.
Intuitives
▶ Be sure that all reasonable alternatives are raised and that each is thoroughly examined.
▶ Listen carefully to the facts that are of concern to others, but don't get bogged down. Keep your creativity engaged.
▶ Let your imagination show others how the team may actually be able to take advantage of a change or problem to produce a better result.
Thinkers
▶ Use your ability for rational analysis and objective criteria to assist in identifying solutions or needed changes.
▶ Avoid personalizing a problem—attack the problem, not the person.
▶ Rephrase the issues being discussed to help the team stay focused and to provide precision and clarity.

How to use your preferences when confronting changes or problems	
Feelers	▶ Remind others on the team to consider the impact of a solution or change on other people.
	▶ As others push for closure, be sure that closure is not reached at the expense of group harmony. Ask if all can live with a proposed solution.
	▶ Make sure everyone gets a chance to speak and be heard.
Judgers	▶ Keep the process goal oriented, and keep a clear agenda in front of the group.
	▶ If the discussion begins to drift, remind the team that, although the discussion may be interesting, it is not relevant to the issue at hand.
	▶ Make sure that there is agreement on how the implementation of changes will be tracked and adjusted.
Perceivers	▶ Help the team avoid rushing to a conclusion.
	▶ Even if it is appealing to the group to make a decision and move on, be sure that the team has the chance for any discussion that may save time and effort in the long run. But be aware of when the discussion has reached the point of diminishing returns.

Individuals need to balance their preferred styles with their less-preferred styles when dealing with problems. This is because the problem-solving process is often blocked by poor communication that results when individuals remain firmly in the grip of only their preferred styles. As a mediator told us, when two parties agree on what the problem is, his work is just about complete. Without effective communication, the chances of agreeing on the problem will diminish greatly. When you consciously call upon your less-preferred style, you often get a different perspective on the problem. The communication process will be improved by each person altering his/her natural preferences. However, once the issue has been agreed upon and understood thoroughly, it becomes a matter of generating good ideas by using the strengths of the preferred style.

The project schedule and budget will be impacted by a change, yet often no one bothers to actually change the schedule or budget. No one (especially a Judger) likes to admit that a change is needed or go through the hassle of altering the project plan. Another common reason is that of unfounded optimism that the change will have no effect, so there is no reason to change the plan. Even if the completion date isn't affected, individual task schedules and personnel assignments can be. Changes must be incorporated into the project plan, or any hope of control or accurate monitoring of progress is lost. Documentation of the change and its anticipated effect should be made when the change is initiated—not later. It becomes very difficult to sort out the effects after the fact. It is prudent to build project team awareness at the start of the project about the importance of incorporating changes in the project plan.

If the scope and budget need to be changed, get the team members to come up with the revisions. Forcing revisions on the team is counterproductive because they are less likely to buy into the changes if they are not involved in developing them.

There are some unplanned, unanticipated project changes that can change the course and outcome of the project. Be open to them. Don't let the project plan become a rigid impediment. The key to the huge success of Post-It Notes is the semisticky adhesive used on the backs of the notes; yet, this adhesive was the result of a "failed" attempt to develop a strong adhesive. Fortunately for the 3M Company, the project manager had the flexibility to recognize that it was time to abandon the original plan and work on putting the newly discovered adhesive to another marketable use. Consider the Apollo 13 mission. Clearly, the planned and desired outcome of the mission was to put the crew on the moon and return them to Earth. When the spacecraft's oxygen system exploded en route to the moon, the mission suddenly changed to getting the crew home safely. To do so, the entire project team had to come to the realization that the mission to the moon had to be abandoned. They had to change their mindsets to deal with problems they had never anticipated in very limited time to avoid the loss of the crew. Had the team adhered to its standard approaches and resisted changing the original plan, the results would have been disastrous.

Dealing with Scope Changes

No project ever goes according to plan!

Dealing effectively with scope changes provides an opportunity to demonstrate leadership. Failing to effectively deal with scope changes will act to the detriment of the team, the project, and the customer's satisfaction. Even when the change in scope results from directions from the customer, some project managers are reluctant to approach the customer with a request for added time or budget to deal with the changed scope. When a customer makes a request for a work product, ask yourself the following:

- ☐ Does this product fall outside the scope of work?
- ☐ Will this impact any other parts of our scope, budget, or schedule?

If the answer to either of the above is yes, it is time to gain the customer's approval and authorization *before* proceeding with the additional work.

Sometimes events beyond the control of either the customer or the project team result in a scope change, and still many project managers are reluctant to approach the customer. Check any of the following top-ten reasons (they are actually excuses) that you have used to avoid asking for a scope change, either in a timely way or not at all.

- ☐ It's too late to mention it.
- ☐ If I take time to work out a scope change now, I'll miss the project deadline.
- ☐ It's a gray area and I can't really justify it.
- ☐ The paperwork is too much hassle.

☐ It's too small to ask for.

☐ Look at it as marketing investment for more work in the future.

☐ Another task is under budget, so there is no need to worry about it.

☐ I'll tell the customer later; I'm too busy now.

☐ We'll make it up in the next phase.

☐ There's no way we are going to spend this whole budget anyway, so why ask for a change? (Typically said when a change occurs in the early stages of a project!)

Figure 6-4 appears on page 149
Many project managers fear that raising a scope change with the customer will be confrontational and may adversely affect the relationship with a customer. Ironically, what we find from interviewing many customers about satisfaction with service providers is that a leading cause of dissatisfaction is failure to promptly report a scope change before additional work is done. A project leader knows that waiting until later to tell the customer that more budget or time is needed—because work related to a scope change already has been performed—will cause a bigger confrontation and greater stress on the relationship than will addressing the change with the customer at the time it occurs. Your customers will be upset if you tell them there will be extra costs or time needed, after you have already done the extra work. A project leader steps forward at the time the potential scope change occurs and discusses it with the customer before proceeding with the additional work. Figure 6-4 presents a checklist of steps to help you deal with scope changes.

Dealing with Conflict

Situations involving conflict among team members or between the team and the customer can arise on projects for a variety of reasons. Although there is a common tendency to view conflict as a threat to team and customer relationships, effectively managed conflict can be constructive because it can lead to the following:

▶ Clarification and greater understanding

▶ Identification of alternative solutions

▶ Challenging of ideas or actions that have been insufficiently reasoned

▶ Production of a better solution

▶ Building of cohesiveness in the team or with the customers

Conflicts can arise from the following:

▶ Factors related to the *Content, Procedural,* or *Relationship* sides of the Triangle of Needs

▶ Differences in personal styles

▶ Differences in personal values

When conflicts arise, understanding which of the factors listed above is the underlying cause is an important element in achieving a positive result from the conflict.

Content-Related Causes

Factors on the Content side of the triangle often relate to differing perceptions of the facts and/or the problem the parties are trying to address. Experienced mediators say that when people agree on what actually is the problem, they are then in the position to develop solutions. Often, people get stuck arguing over the definition of the problem. For example, is the problem with an overcrowded highway the lack of an adequate number of lanes, or is it that not enough is being done to encourage carpooling? One perception of the problem leads to adding lanes to accommodate traffic, while another leads to converting a lane to carpool/HOV to force people to share rides and reduce traffic. A solution will not be found until there is agreement on the problem. Data problems also often show up on the Content side of the triangle. We have seen many instances where different individuals had differing opinions about the amount of data needed or about the quality of the data collected. Differing analysis of the same data by different people also can lead to conflict.

Procedural-Related Causes

Content conflicts can escalate if the project process is also flawed. For example, if the project design isn't conducted in a logical and progressive order, then conflicts between disciplines will occur. If it is not clear how decisions will be made, or if there is no attempt to balance team or stakeholder participation, then conflicts may occur if people feel excluded or ignored. Some may choose not to speak up if they feel their communication skills are not as good as others' and, if a procedure is not in place to draw out their concerns, conflicts related to their unexpressed issues may escalate unknowingly.

Relationship-Related Causes

People may come into projects carrying prior relationship baggage with others. It is very difficult for people to set aside previous bad experiences with one another. Be aware of the interpersonal history of people on your teams and with project stakeholders.

▶ *Style-Related Causes.* Differences in personal style preferences also influence the potential for conflict. Styles affect behavior related to dealing with conflict in the following ways:

Communication	
Extraverts	*Introverts*
• Tend to address conflict head on • Are energized by verbal exchange—vent and let go • Are surprised and irritated when an Introvert erupts later over an exchange the Extravert considers to be over	• Tend to become quiet and retreat to collect their thoughts when faced with conflict • Are drained by verbal exchange—may stew and erupt later

Level of detail	
Sensors	*Intuitives*
• Consider details very important in resolving an issue • Tend to bring the process to a halt by bringing up more details if they disagree on an issue	• Tend to gloss over the details and treat related issues as minor • Want to resolve issues in the context of how the issues affect the big picture and consider discussion of details as holding up progress

Decision making	
Thinkers	*Feelers*
• Can be irritated when others view their use of impersonal, objective criteria as unsympathetic or judgmental • Tend to discount people's feelings when making a decision	• Can be irritated when others do not use personal, value-related criteria in making decisions • May oppose a decision because of the way the people involved are handled

Process and structure	
Judgers	*Perceivers*
• Tend to resist changes in the project plan • Want to get the issue resolved and move on	• Are irritated by perceived rigidity or lack of flexibility • Want to talk about options and be sure all alternatives are considered before moving on

▶ *Value-Related Causes.* Individuals have different values, and these differing values will cause conflict if someone feels his or her values are being violated. These value differences can be among the most difficult to resolve. Remember, the conflicts over the spotted owl issue in the northwest United States? The basic values of the conservationists and the logging communities were not going to change, making this a very emotional and difficult conflict. When value differences contribute to a conflict, recognize that it is pointless to try to change basic values. Instead, look for alternatives to satisfy these basic values within the context of solving the problem.

One example of a conflict resolution approach that satisfied conflicting basic values involved a water supply reservoir project in the eastern United States. The planners of the project first approached the water quality agency in the state, who told them that their permit would require that all trees be removed from the proposed reservoir site to protect against leaching of organic materials into the water, which would adversely affect water quality. Conversely, when the planners approached the fish and game agency, they were told that their permit would require that the trees *remain* in the reservoir to provide fish habitat. Clearly, neither of the agencies was going to give up its values of "clean water" and "fish habitat." The planners considered the conflict in values and came up with a solution that addressed both values. The planners proposed providing shallow areas around the perimeter of the reservoir where vegetation would be provided for fish habitat and removing the trees from the deeper parts of the reservoir to protect water quality. This approach was acceptable to both agencies. Then the state health agency said that there could be no shallow pools around the perimeter because of their concern about mosquito propagation in the shallow areas. The health agency's value of "no mosquito propagation" was in conflict with the value of providing "fish habitat." The planners ultimately proposed to stock the shallow areas with mosquito fish, a proposal that resolved the last of the conflicting values. The potential conflict in underlying values was resolved by creatively searching for win-win solutions rather than trying to deal with negotiating a win-lose compromise .

The Importance of Attitude in Resolving Conflicts

A successful attitude is essential to dealing with conflict in a constructive way. Such an attitude is made up of the following components:

- ► You want to resolve the conflict.
- ► You want to maintain the relationship.
- ► You are willing to take initiative.
- ► You do not need to be right.
- ► You will do what it takes. The satisfaction is in the resolution.
- ► You accept that the other person may not feel the same way.

It is important to recognize the things you can change to manage conflict and those—such as values—you cannot. For example, once we understand that conflict is being caused by a shortcoming with a procedure, we can seek a change in the procedure that will address the perceived shortcoming. Similarly, we can make changes in a Content issue, such as adjusting the data collection plan, once we realize that this is the underlying cause of a conflict. This is not to say that it will always be practical to make the change. For example, in some cases it may not be practical to ever collect enough data to satisfy a perfectionist. But, by understanding that data collection is the conflict cause, it will be possible to address the concern, demonstrate the time and cost implications, and make changes that are affordable and that will ease the conflict by demonstrating a willingness to do all that is possible to address the concern.

Personal styles and relationships are causes that we can choose to make changes to address. But unless we choose to do so and exert the substantial personal energy required, these causes will fall into the unchangeable category, making conflict management difficult if not impossible. Basic values are, as discussed previously, unchangeable. Accept this fact and look for alternatives to satisfy the basic values within the context of solving the problem.

Status Reports

All three sides of the Triangle of Needs need to be assessed and balanced in relation to project status and any status reporting, as follows:

Content concerns:
▶ Schedule status
▶ Budget status
▶ Cost/effort to complete the project
▶ Availability of resources

Procedural concerns:
▶ Project reviews being conducted and acted upon
▶ Project status reporting

Relationship concerns:
▶ How do the customer and stakeholders feel about the project?
▶ How do team members feel about the project?
▶ Are there any areas of conflict or misunderstanding?

Effective progress reports are also one way to keep your customer and your management from distracting you with frequent inquiries as to what's happening on the project. Keep the reports as brief as possible, using pictures or simple graphics whenever possible. In deciding on a report format, put yourself in your customer's shoes. What are they really interested in knowing? Adjust your style of presentation to match the customer's style preferences. For example, if your customer is an Intuitive, don't fill the status report with details. If the customer is a Sensor, be sure to include all of the details of interest to him or her.

In dealing with a city council that was predominantly made up of Intuitives, we used a one-page monthly status report for a $27,000,000 municipal construction project. The report consisted of an aerial photograph of the project taken at the first of the month, a graph presenting the earned value versus budgeted cost, and a one-half page list of significant accomplishments and issues. The report was very effective. The photograph conveyed a tremendous amount of information in one glance, and appealed to the preference of the city council to see the big picture.

In another case, we interviewed one of our client's customers as part of a study to determine their satisfaction with our client's services. In this project, our client's project manager was a Sensor and his customer was an Intuitive. When we asked about status report, the customer became very animated and angry. He said he had told the project manager that he did not want the two-inch-thick progress reports that he was receiving because they contained details of no interest to him—but the two-inch-thick reports kept coming. Obviously, the project manager was preparing reports that appealed to himself without adjusting to the customer's style preferences. It was also apparent that the project manager needed to work on his listening skills!

The frequency of reports depends on the duration of the project. Obviously, reports can be much less frequent on a ten-year-duration project (quarterly will probably do), as compared to a ten-month project (monthly will be needed).

Exercises

1. Have someone ask you the questions on the status checklist (Figure 6-3) for a current project or task you are involved in and ask you follow-up questions as appropriate. After having this discussion, address the following questions:

    a. Did you realize you knew less about the project than you thought you did before the discussion? What were the gaps in your knowledge?

    b. What action items did you develop?

    c. How long did the discussion take? Was it worth the time?

2. Identify and make any needed changes to the status checklist (Figure 6-3) so that you can use it to structure a discussion about work status with a task leader or key team member. Have the discussion. After having this discussion, address the following questions:

    a. Do you know more about the project than you did before the discussion?

    b. What action items did you develop?

    c. How long did the discussion take? Was it worth the time?

    d. Did you see any relationship between your personal style preferences and the discussion that you had?

3. Working with your team, pick a previous project with a scope change that became an issue. Discuss how it could have been handled more effectively as it relates to the following:

    a. Planning

    b. Personal style preferences of those involved (your team and the customer)

    c. Procedures (or lack thereof) for addressing scope changes

    d. Failure to address one or more of the sides of the Triangle of Needs

4. Think of a situation with a team member or customer where a conflict exists and what the underlying cause may be:

    a. Factors related to the *Content, Procedural,* or *Relationship* sides of the Triangle of Needs

    b. Differences in personal styles

    c. Differences in personal values

What changes could you make to resolve the conflict? What changes are you willing to make to resolve the conflict? If the conflict is related to differences in personal values, what are these differences and what solutions could resolve the conflict while preserving both parties' values?

Figure 6-1   *Checklist for Effective Project Communication*

DOs

- [ ] Define the protocols for normal and unusual communications with:
  - ▶ Your customer
  - ▶ Team members
  - ▶ Subcontractors (if applicable)
- [ ] Include protocol elements:
  - ▶ Signature authorities
  - ▶ Communications channels (i.e., who is authorized to talk to whom)
  - ▶ Documentation procedures for telephone conversations
  - ▶ Memoranda
  - ▶ Team meetings (schedule and standard agenda)
  - ▶ Standard routing list
  - ▶ Confidentiality provisions
  - ▶ Dealing with the media
- [ ] Verify all mail addresses, e-mail addresses, and phone numbers of the customer and subcontractors.
- [ ] Give clear work assignments to every project team member, including scope, budget, and schedule. Show them how their assignments fit into the big picture of successfully completing the project. Describe complete documentation requirements to make the task product reviewable. Encourage team members to communicate with you about potential problems at the earliest possible time. Make it safe to be the messenger of bad news.
- [ ] Schedule administrative and technical support services as soon as you know when an item is promised to the customer (review drafts, portions of a report or contract, etc.)
- [ ] Develop and use a mechanism for communicating changes in the scope, budget, and schedule.
- [ ] Conduct a kickoff meeting. Consider use of a facilitator if the project team includes more than ten people or more than four parties with potentially divergent interests.
- [ ] Develop and circulate an agenda for the kickoff meeting at least two days prior to the meeting.
- [ ] Invite the appropriate personnel (normally, the entire team) to the kickoff meeting.
- [ ] Hold regular team meetings. Define meeting frequency in the kickoff meeting.
- [ ] Recognize people's tendencies to avoid conflict, and develop a communications plan that promotes open communication.
- [ ] Recognize the tendency to be overoptimistic about meeting schedules and budgets; assume that there will be interruptions, delays, and unforeseen costs.
- [ ] Communicate schedule changes to the project team regularly, including all support staff.
- [ ] Hold a closeout meeting to get feedback, share lessons learned, and collect important budget and scheduling information for future estimating purposes (see Chapter Seven).
- [ ] Be sensitive to team members' personal style preferences. Adjust your preferences as appropriate.

DON'Ts

- [ ] Assume things are going well just because you have not been approached with problems.
- [ ] Assume your project is no. 1 priority for staff resources.
- [ ] Wait until a document is ready to produce to schedule support services.
- [ ] Give last-minute or rush tail-end tasks without warning or discussion on what is a feasible work product in the allowable time frame.
- [ ] Assume that team members' personal style preferences match yours.

Figure 6-2   *Checklist of Questions to Team Members to Solicit Project Status Information*

☐ Do you need any resources that you don't have?

☐ Do you know of anything that will cause schedule or budget problems?

☐ Any chance you'll finish early? Under budget? Could you explain the reasons we're doing so well so we can use the information on other projects?

☐ What's the most difficult aspect of your work?

☐ What problems do you anticipate? What kinds of steps should we take?

☐ What's your assessment of the overall project status?

☐ What should we be doing differently?

☐ What percent complete is your work? What's the basis for your estimate?

☐ How much effort and time do you estimate it will take to finish your work? What's the basis for your estimate?

☐ What problems are you having interfacing with other project tasks?

☐ Which incomplete tasks are in progress? Which ones haven't started?

☐ What other demands are there on your time?

☐ How satisfied is the customer with the project at this stage? How do you know?

Figure 6-3   *Project Review Checklist*

Date _____

Project Name _____

1. Does the authorized scope of services include all work that has already been conducted or that is currently planned (i.e., has there been scope creep)?

   Yes   No   (circle one)

   Explain:

2. If out-of-scope work has been or is expected to be performed, have we requested, in writing, additional funding and time for all out-of-scope items?

   Yes   No

   Amount of request (hrs): _____ Amount of request ($): _____

   Date approval received: _____

3. Is the current staffing adequate to complete the work as planned?

   Yes   No

   Discuss (availability of personnel, particular requirements, etc.):

4. Are all members of the project team (including support staff) fully informed of the project schedule?

   Yes   No

   How were they informed and date?

5. Are all consultants and subconsultants on track with their scopes, budgets, and schedules?

   Yes   No   Not applicable

   What problems, if any, currently exist?

   Do you anticipate any problems? Explain:

6. Have all quality control measures planned to occur to date actually occurred?

   Yes   No

   If no, please discuss:

7. Does the project team have any conflicts/current problems affecting this project?

   Yes   No

   If yes, please discuss:

continued on next page

Figure 6-3   *Project Review Checklist (continued)*

8. Is the existing budget expected to be adequate to complete the work?

    Yes   No

    Staff hours remaining: _____ Staff hours required to complete: _____

9. Have all invoices to-date been sent or received?

    Yes   No

10. Have all invoices over 90 days old been paid?

    Yes   No

    Amount of outstanding invoices: $ _____

11. Is the customer satisfied with the project to date?

    Yes   No

    How do you know?

12. What is your next planned contact with your customer (internal or external)?

13. Has public/stakeholder involvement, if applicable, been successful so far?

    Yes   No

    How do you know?

14. Discussion of action items, responsibility, and implementation schedule.

Action #	Action Description	By Whom	By When

Figure 6-4   *Checklist for Scope Creep Control*

☐ At the start of the project, set up a procedure for dealing with out-of-scope work. Agree with the customer on a form and procedure for advising the customer when potential scope changes occur. This educates the customer that scope changes may occur. It is also much easier to agree on a procedure for handling changes at the start of the project. During the middle of a project, it may be very stressful to work out a procedure at the same time you are trying to accomplish the work and get agreement upon a change the customer was not expecting.

☐ Provide a project guide/manual to all team members so they know the procedure for changes in scope and schedule. Include a discussion of ongoing work that may be out of scope at each project team meeting, and agree on the immediate action steps to discuss the change with the customer if such work is occurring or is about to occur.

☐ Reduce the chances of your own team creating out-of-scope work by making sure team members know the scope, budget, and schedule for their tasks. Put the scope in the project guide/management plan, and monitor ongoing work so that any deviation from the scope can be quickly controlled.

☐ Set up a separate account in your project cost accounting system to track the costs associated with out-of-scope additional work.

☐ When a customer makes a request for a work product, ask yourself the following:

    ☐ Is this within the scope?

    ☐ Do I need anyone else's input before I agree to this?

    ☐ Will this impact any other parts of our scope or schedule?

☐ Discuss out-of-scope work with customer and document their agreement on the effects on budget and schedule before the work is started. Pending completion of a formal scope amendment, promptly send an e-mail or fax to the customer, describing your understanding of the directive. This will confirm the understanding that the directive represents a change in scope, assure the customer that work is proceeding per the customer's direction, and describe the effects on schedule and budget.

☐ Document all internal and external communications on out-of-scope work, especially verbal directives from the customer.

CHAPTER

SEVEN

Have We Crossed the Finish Line?
Closing the Project

Addressing the Triangle of Needs

Are there boxes of files from completed projects stacked in your office that you have been waiting to have some time to properly sort and file? These boxes are a sign that you stopped short of effectively completing the project. The winning team never stops a few feet short of the finish line. When the finish line is in sight, a winning lead dog does not slow down or stop for a rest. Yet, many project teams start to lose their enthusiasm and focus as the project winds down and, as a result, do not effectively complete the project. A project leader keeps the team focused until the project is completed and sees that project closeout addresses all sides of the Triangle of Needs:

- ▶ *Budget.* Care is taken to see that the project budget is not blown in the last 10% of a project, a common problem. Costs for closeout tasks are budgeted and time is scheduled so that the closeout tasks are thoroughly completed in a timely way.
- ▶ *Technical transfer.* Project information is summarized and made readily available so that other project teams can learn from your experiences.
- ▶ *Tying up the loose ends.* All of the "little" project items that seem insignificant in themselves are addressed to avoid the loose ends that may irritate the customer to the exclusion of the 99.5% of the project that went really well.

- ▶ *Filing.* Critical information is placed in a central project file and unneeded materials are removed. Materials left in files can be damaging in the event of project problems that later lead to litigation. In one legal case where we were examining documents from the defendant's files, we discovered a memo from the firm's technical expert in which he wondered how his firm could ever have been stupid enough to hire the incompetent people who did the poor-quality work for which they were being sued. Writing such a memo in the first place was ill-advised; but, failing to sort through the files at project completion and failing to remove the memo was a disaster for the firm. Failing to put critical information in the central project file can also be a problem. A structural problem occurred on a project when a constructed concrete wall began to fail. There was a debate about whether it was a construction or design problem. The individual who did the design work had left the employ of the design firm. His design calculations were not in the central project file and could not be found. The lack of design information placed the design firm at a serious disadvantage in the litigation. The firm eventually participated in a large financial settlement and lost a longtime client in the process.

▶ *Accounting.* All financial information is completed and final costs compiled and compared to budget. Reasons for variations from the planned budget are identified and documented.

▶ *Public relations/marketing.* Concise information about the project is prepared for future public relations or future proposals. Even though project descriptions may have been prepared at project initiation, changes as the project proceeded will have made that information inaccurate. It is all too common to see consulting firms scrambling for information on past projects as they attempt to put together proposals. Public agencies may be scrambling for accurate, up-to-date project information as they prepare annual reports or respond to inquiries. There is no better time than at project completion to see that accurate information is assembled and made accessible to all.

▶ *Team morale.* Team morale will be higher if the team sees a finish to the project rather than seeing loose ends drag on for weeks.

▶ *Customer relationships.* Efficient wrap-up of a project will leave a positive image with your customer. This can lead to follow-on work and good references.

Effective project completion can be the difference between success and failure because project budgets are often blown in the last 10% of a project. When the completion is poorly handled, the customer can become upset and the learning that others could achieve from the project is delayed. Your customer will better remember the most recent events; so, the way you close out the project can have a substantial effect on your reputation and chances for follow-on work.

Human Factors

Most project managers have a relatively easy time understanding the technical and mental chores of finalizing project reports, assembling the final cost information, finishing any tasks, and disposing of any project hardware; but project leaders also understand the emotional issues that arise as project completion approaches. To address these issues, a project leader considers and deals with the needs and personal style preferences of each team member on an individual basis. The following issues are typically encountered:

- ► *Loss of interest.* As the job nears completion, the creative aspects are replaced by more mundane, detail-oriented, administrative-type duties. Intuitives may become bored and disenchanted, and look upon the remaining work as drudgery. If not removed from the project, they may get little done because they will divert their energy to more interesting tasks or in seeking another assignment.

- ► *Anxiety over the future.* If there has been no specific future assignment, a person's anxiety may lead to foot-dragging. The slowdown may result from a conscious decision to spend time looking for another project, or even another job, or from a subconscious desire to stretch out the remaining work to provide security. If the next assignment is viewed as less attractive, team members will not be eager to move on. If the next assignment is not made, unproductive time may continue to be charged to the old project, causing a budget overrun. Those with a preference for Judging may become especially anxious if the lack of definition of their next assignment frustrates their desire for a structured future.

- ► *Loss of enthusiasm.* In the height of an exciting project, team members are bound together by a common mission that energizes them. As the project nears completion, team objectives become less important and individual concerns take over. The remaining work may consist largely of unexciting details, and enthusiasm drops.

- ► *Lack of focus.* Work on more interesting new assignments can disrupt the focus of the team. Their energy soon becomes scattered between the old and new projects, with the old one losing out.

To effectively close out the project, restructure the project team and treat project closeout as a separate "mini-project," with its own plan, budget, and schedule. Focus on the emotional issues of the team members before proceeding with project completion. Consider the personal style preferences and skills of each team member. You may be better served by retaining those with a Sensing preference, who deal well with details, and reassigning the Intuitives to the planning phase of the next project. Those with a Judging preference will likely be more driven to see that project closure is achieved than those with a Perceiving preference. Your positive actions in reassigning and restructuring the remaining team members can calm anxieties and reduce the energy that would otherwise be lost to discussing rumors about the future. Even if reassignment is not possible for everyone on the team, an honest, open approach is much more positive than inaction that can leave everyone on edge. Treating all team members honestly and openly by discussing information about their next assignments demonstrates your respect for the individuals on your team and can encourage those remaining to close out the project.

To effectively engage the remaining team, define the objectives of the close-out effort as though it were a project in itself. Get the team involved in defining these objectives, setting a schedule, and dividing up assignments. Keep communications going through informal staff meetings and frequent, individual contact with each team member. Personally demonstrate your sense of urgency and enthusiasm. Be generous with pats on the back. Recognize the unrewarding nature of this phase of the project. By taking the same pride in completing the project as you did in executing the project, you'll set an example for the team to emulate. The time required for project closeout will be condensed so that you can all move on to the next project and take pride in having efficiently completed the last project. Have a celebration at the completion of the project to: (1) provide recognition, (2) acknowledge team effort and spirit, (3) provide a sense of completion, and (4) set a precedent that will give future teams something to look forward to.

Closeout Checklist

Figure 7-1 appears on page 157

Figure 7-1 is a checklist for project closeout. The checklist contains some items that will apply only to projects funded by an external client that has been invoiced for work on the project. The checklist can be readily modified for other types of projects.

Project Closeout Debriefing Meetings

Figure 7-2 appears on page 158; Figure 7-3 appears on page 159

You and your organization can learn by observing the causes and consequences (both good and bad) of your actions on the project and by make conclusions about how to act differently on future projects or reinforce and build on effective actions taken. A project closeout debriefing meeting can provide great dividends for a relatively small investment of time. The project leader can prepare for such a meeting by gathering and summarizing information on actual versus planned scope, schedule and budget performance, customer comments, quality control comments, and his/her own observations. Figure 7-2 presents an example agenda for a debriefing meeting attended by the project leader and the team members. Figure 7-3 lists sample questions to address in the closeout meeting, as well as an outline for the summary of the meeting.

Make the closeout meeting a postproject success acknowledgment as well. With the right attitude, this session can be fun. Create some fun team awards voted on by peers, such as "Person in least need of sleep," "Person with most creative ideas." Even if the project didn't go well, some ceremony can help the team reach closure. We know of one team whose project had not gone well. To bring the project to closure and move on to the next one, the team leader arranged a pizza party during which the team took a copy of the project report to the back lot of their property, dug a grave, and buried the report.

Figure 7-4 appears on page 160 A project debriefing with the customer can provide very valuable information that can be used to improve team performance. Figure 7-4 presents some suggested questions that can be used in a customer debriefing. It may be useful to have someone other than the day-to-day project customer contact conduct the customer debriefings. The customer may be more frank and open with another person. As project leader, you should share the customer's feedback with the project team and focus on what was learned and how that can be used to improve team performance.

This will bring you full circle and close the loop on meeting your customer's, stakeholders', and team's expectations. It also helps keep your Triangle of Needs balanced and in place, and provides a strong foundation for future or ongoing projects.

We all learn from every project and will become stronger and wiser leaders if we view all of our experiences (good and bad) as "learning opportunities." Just as dogsled teams don't rest on their laurels for long after each race—but noisily express their excitement as they wait to be harnessed for the next practice run—so too can you! We highly recommend you take your team on "practice runs" using all of the checklists and exercises in each chapter. And remember to

M ake

U nprecedented

S uccess

H appen

Exercises

1. Fill out the project closeout checklist (Figure 7-1) to a current project and assign the team members to the closeout activities. Distribute and discuss the checklist at the next team meeting and agree on a schedule to carry out the activities. Follow up with the team during the closeout phase of the project. How many of the items on the list have been done? What were the causes for items that are not complete? Complete the remaining items on the list.

2. Complete Figure 7-3 for a current, ongoing project pretending that it is now done. How is your project doing? What could you do to improve it for completion?

3. Complete Figure 7-4, putting yourself in your customer's shoes on a current project. What are you doing well, and what do you need to adjust to have a more satisfied customer?

4. Interview a customer about a project that your team recently completed using the customer debriefing questionnaire (Figure 7-4) as a guide. Discuss the results with the project team and work with your team to identify how your service can be improved in the future. Also identify those aspects that were particularly successful so that they can be reinforced for use on the next project.

5. Using Figure 7-2 as a starting point, develop an agenda for a project closeout meeting that is nearing project completion, and hold the meeting at project completion. Pay particular attention to making the celebration aspect significant. Were there lessons learned on the project that will be helpful to other teams? Use Figure 7-3 as a guide for preparing notes that document these lessons learned and see that the information is transferred to members of your and other teams.

Figure 7-1   *Project Closeout Checklist*

Documentation	Yes	No	Assigned to
Project Manager's personal project files in central project file (CPF)?			
Team members' personal project files in CPF?			
Memos, letters, etc. in correct order by date in CPF?			
Latest version of computer disks and hard copies in CPF and indexed?			
Calculations, test results, final contract documents in CPF?			
List of tasks and persons who completed them in CPF?			
Obsolete reports, calculations, memos, etc. thrown away?			
Master index for CPF developed?			
CPF numbered and in accessible storage?			

Accounting	Yes	No	Assigned to
Final budget changes documented?			
Final invoice paid?			
Project accounting closed out?			
Final costs reviewed and placed in CPF?			
Overall project costs determined and analyzed for lessons learned?			
Final invoices received from all subconsultants and paid?			

Communication	Yes	No	Assigned to
Team debrief meeting held?			
Lessons learned from team documented and circulated?			
Customer debrief meeting held?			
Lessons learned from customer documented and circulated?			
Any potential remaining issues? (attach documentation)			
Project team recognition/celebration event held?			

Marketing/public relations	Yes	No	Assigned to
Have the following been sent to marketing/public relations:			
• Accurate project descriptions			
• Project team members and related tasks			
• Customer reference, address, and phone number			
• Relevant project photos and videos			

Figure 7-2   *Example Agenda For Project Debriefing Meeting*

What	How	Who	Time
Introduction, purpose, agenda, roles, guidelines	Review and agree	Project leader as facilitator	0–5 minutes
Scope and schedule performance	Review actual versus planned and causes for differences	Project leader as facilitator	25–45 minutes
Budget performance	Review actual versus planned and causes for differences	Project leader as facilitator	5–30 minutes
Lessons learned	Respond to list of questions in Figure 7-3	Project leader as facilitator	40–80 minutes
Celebration	Team award, MVP, pizza	All	15–30 minutes
Total			1.5–3 hours

Figure 7-3 *Project Debriefing Summary*

1. What went well that we would do the same way next time? What aspects were particularly enjoyable?

2. What would we do differently next time?

3. What improvements should we make regarding staffing, technical support, and communications?

4. What scope changes occurred, and how did we deal with them?

5. What was the quality of the final project deliverable?

6. Were any new tools or procedures developed that could be used on other projects?

7. On a scale of 1 to 5 (5 being best), how do you rate the Relationship side of the Triangle of Needs between:

   a. Customer and team _____

   b. Team and leader _____

   c. Team and stakeholders _____

   How did styles influence the relationship and interactions?

   What would have improved the ratings by 1 point? By 2 points?

8. Was the customer satisfied with the end product? What did the customer like or dislike?

Distribution list for these notes:

Figure 7-4   *Customer Debriefing Questionnaire*

Person Interviewed_____ By_____ Date_____

General

What types of services did we perform for you?

Who has been your primary contact?

Overall service

What is the quality of your work experience with our team?

What are the strengths of our team?

How could we improve our service to you?

Who are our competitors, and how do they compare?

Negotiations

How do you rate our understanding of your needs before we began work?

Are we reasonable to negotiate with?

How do our project costs compare to competitors?

Production

How well did we keep you informed as the project proceeded?

How responsive were we? Did our team return calls quickly? Were your questions answered appropriately?

Did the right people work on your project?

How well did we listen to you?

How well did we meet schedules?

How do you rate our technical competence?

How do you rate the quality of our work products?

How well did we perform relative to budgets?

Did we advise you of scope changes in a timely way?

Did you receive value proportional to your costs?

Billing (for a revenue-producing project for which a customer was billed)

Did our invoices contain the right amount of detail?

Were invoices timely?

Were invoices accurate?

Were questions about invoices handled in a timely way and appropriately?

Follow-up

Did our team contact you after the project was complete to see if you were satisfied?

Do you think we are interested in providing other services to you?

Would you hire us again?

Are there any specific future projects for which you are considering or would consider hiring us?

Closing

Is there anything I should have asked you but didn't?

If you had one piece of advice to give us, what would it be?

References

Bennis, W., and Goldsmith, J. (2003). *Learning to Lead: A Workbook on Becoming a Leader.* Basic Books, Perseus Books Group, Cambridge, Mass.

Blaylock, B.K. (1983). "Teamwork in a Simulated Production Environment." *Journal of Psychological Type*, 6, p. 32–37.

Brock, S.A. (1994). *Using Type in Selling.* Consulting Psychologists Press, Inc., Palo Alto, Calif.

Campbell, D. (2002). *Campbell Leadership Descriptor, Participant Workbook.* Jossey-Bass Pfeiffer, San Francisco.

Carnegie, D. and Associates. (1993). *The Leader in You.* Simon & Schuster, New York.

Cary, B., and de Marcken, G. (1999). *Born to Pull.* Pfeifer-Hamilton Publishers, Duluth, Minn.

Covey, S.R. (1992). *Principle-Centered Leadership.* Fireside Press, New York.

Covey, S.R. (1989). *The Seven Habits of Highly Effective People.* Simon & Schuster, New York.

Covey, S.R. (1999). "Making Time for Gorillas." *Harvard Business Review*, Nov./Dec.

Culp, G.L., and Smith, R.A. (1992). *Managing People (Including Yourself) for Project Success.* Van Nostrand Reinhold, New York.

Culp, G.L., and Smith, R.A. (2001). "Understanding Psychological Type to Improve Project Team Performance." *Journal of Management in Engineering*, Jan./Feb.

DePree, M. (1989). *Leadership Is an Art.* Bantam Doubleday Dell Publishing Group, Inc., New York.

Garfield, C. (1986). *Peak Performers: The New Heroes of American Business.* William Morrow and Co., New York.

Hirsch, Sandra K., and Kise, Jane A.G. (2000). *Introduction to Type and Coaching.* Consulting Psychologists Press, Inc., Palo Alto, Calif.

Kouzes, J.M., and Posner, B.Z. (2002). *The Leadership Challenge,* 3rd Ed. Jossey-Bass, Inc., San Francisco.

Kouzes, J.M., and Posner, B.Z. (2001). *Leadership Practices Inventory, Participants Workbook.* Jossey-Bass Pfeiffer, San Francisco.

Kroeger, O., and Thuesen, J.M. (1992). *Type Talk at Work.* Dell Publishing, New York.

McCauley, C.D., Moxley, R.S., and Van Velsor, E. (1998). *The Center for Creative Leadership Development Handbook of Leadership Development.* Jossey-Bass Publishers, San Francisco.

Mackenzie, R. A. (1972). *The Time Trap.* McGraw-Hill, New York.

Moore, C.W. (1986). *The Mediation Process.* Center for Dispute Resolution Associates, Boulder, Colo.

Myers, I.B. (1998). *Introduction to Type.* Consulting Psychologists Press, Inc., Palo Alto, Calif.

Myers, I.B., McCauley, M.H., Quenk, N.L., and Hammer, A.L. (1998). *MBTI Manual: A Guide to the Development and Use of the Myers-Briggs Type Indicator.* Consulting Psychologists Press, Inc., Palo Alto, Calif.

Myers, K.D., and Kirby, L.K. (1994). *Introduction to Type Dynamics and Development.* Consulting Psychologists Press, Inc., Palo Alto, Calif.

Nilsen, D., and Campbell, D. (1998). *Developmental Planning Guide for the Campbell Leadership Index.* NCS Pearson, Minneapolis, Minn.

Robbins, S.P. (2002). *The Truth about Managing People.* Prentice Hall, Upper Saddle River, N.J.

Rogers, C.R. (1980). *A Way of Being.* Houghton Mifflin, Boston.

Waitley, D. (1983). *Seeds of Greatness.* Simon & Schuster, New York.

Zemke, R., and Anderson, K. (1997). *Coaching Knock Your Socks Off Service.* AMACON, New York.

Index